Catalytic Conversion of Energy Resources into High Value-Added Products

Catalytic Conversion of Energy Resources into High Value-Added Products

Editors

José Luis Pinilla
Isabel Suelves
Tomás García

MDPI • Basel • Beijing • Wuhan • Barcelona • Belgrade • Manchester • Tokyo • Cluj • Tianjin

Editors
José Luis Pinilla
ICB-CSIC, Institute of
Carbochemistry, CSIC-Spanish
National Research Council
Spain

Isabel Suelves
ICB-CSIC, Institute of
Carbochemistry, CSIC-Spanish
National Research Council
Spain

Tomás García
ICB-CSIC, Institute of
Carbochemistry, CSIC-Spanish
National Research Council
Spain

Editorial Office
MDPI
St. Alban-Anlage 66
4052 Basel, Switzerland

This is a reprint of articles from the Special Issue published online in the open access journal *Energies* (ISSN 1996-1073) (available at: https://www.mdpi.com/journal/energies/special_issues/Catalytic_Conversion_Energy_Resources).

For citation purposes, cite each article independently as indicated on the article page online and as indicated below:

LastName, A.A.; LastName, B.B.; LastName, C.C. Article Title. *Journal Name* **Year**, *Volume Number*, Page Range.

ISBN 978-3-0365-1074-3 (Hbk)
ISBN 978-3-0365-1075-0 (PDF)

Contents

About the Editors

José Luis Pinilla (PhD in Chemical Engineering) is a Tenured Researcher at the Institute of Carboquímica (ICB), which belongs to the Spanish National Research Council (CSIC). He has developed his scientific career at the ICB-CSIC and at the Department of Chemical Engineering of Imperial College London (UK). His research lines are mainly focused on the hydrogen production from hydrocarbons without CO_2 emissions, the synthesis and application of advanced materials based on nanostructured carbon materials (carbon nanofilaments and graphene derivatives) and the development of different catalytic biomass conversion processes. His overall goal is the development of more efficient and sustainable processes for the conversion of fossil and renewable origin resources into fuels, added value products and energy.

Isabel Suelves (Ph.D.). Dra. Suelves (PhD in Chemistry) is a Scientific Researcher from CSIC at the Instituto de Carboquimica (ICB-CSIC). Since February 2019, she has been the Director of the Institute. During recent years, her research has been focused on the simultaneous production of hydrogen and nanostructured carbon via the catalytic decomposition of hydrocarbons, studying aspects related to the development of catalysts and the optimization of the process conditions, as well as the development of the most suitable technology for the scaling up of the process. Additionally, she has opened new research lines related to the search of applications for the carbonaceous material, especially those related to energy storage or biomass conversion, via the development of advanced catalytic systems.

Tomás García (Ph.D.). Dr. García is a Senior Scientific researcher at the ICB-CSIC. His research is devoted to the development of new preparative methodologies to produce highly active heterogeneous catalysts for different catalytic reactions, mainly cracking and total and selective oxidation. Examples include the development of nanocrystalline metal oxides and supported metal catalysts for the oxidative destruction of Volatile Organic Compounds (VOCs) and the synthesis of bifunctional hierarchical zeolites for bio-oil upgrading. In 2007, he won the Aragon Investiga Award to Young Researchers, sponsored by the regional government, for the excellence of those under 35 years old.

Preface to "Catalytic Conversion of Energy Resources into High Value-Added Products"

Catalysis is involved in a large number of industrial processes for the production of chemicals, pharmaceuticals, fuels, and the cleaning or suppression of environmental pollutants. More than 90% of the chemical processes used in industry use catalysts, with heterogeneous catalysts being the most used (80%). Although the global market for catalysts is important per se, the greatest impact of catalysts comes from the value generated by the chemicals and fuels they produce.

The growing concern about the massive use of fossil fuels and its effect on climate change has encouraged research in the development of technologies related to a more efficient use of energy resources. Scientific advances in this area of knowledge are therefore essential to meet the challenges related to global warming and the finite nature of fossil fuels. Therefore, the development of active, selective and energy efficient heterogeneous catalytic processes is of paramount importance for the production of high-value-added products from energy resources in a more sustainable manner.

In this Special Issue of Energies, we collect some of the latest progress in the development of cleaner, more efficient processes for the conversion of these feedstocks into valuable fuels, chemicals and energy. A total of 8 high-quality papers focused on different catalytic systems are showcased. Most of the works are focused on the conversion of biomass related materials which clearly reflects the paramount importance that the biorefinery concept will play in the years to come.

The novelty and contributions of these papers are briefly summarized in the next paragraphs.

Ramirez-Reina et al address the upgrading of biogas, a gaseous mixture of methane and carbon dioxide produced by the anaerobic digestion of biodegradable matter, via dry reforming into high value syngas using nickel-based catalysts. The novelty of this work lies on the use of multicomponent Ni-Sn/CeO$_2$-Al$_2$O$_3$ catalyst and the optimization of the process conditions, namely temperature and biogas composition. When compared towards a benchmark catalyst, the multicomponent catalyst achieved similar conversion and benefited from greater coke resistance, confirming the promotion effect of both tin and ceria.

One of the most critical technical challenges in biomass gasification is the tars formation, which can cause serious risks to downstream equipment. Therefore, tars should be removed from the biomass gasification effluent stream. In the contribution by Millan et al, a commercial Ni/Al$_2$O$_3$ catalyst was used to determine the optimum conditions of the steam reforming of toluene, used as tar model compound. Thus, the influence of reforming temperature, steam to carbon molar ratio (S/C) and gas hourly space velocity (GHSV) on the toluene reforming performance was thoroughly studied.

These process parameters are considered key to fine-tune the reaction and maximize the overall performance. A temperature of 800 °C, GHSV of 61,200 h^{-1} and S/C ratio of 3 provided the most suitable reaction conditions for toluene conversion and H$_2$ production in steam reforming of toluene. Under these optimized conditions, a steady state of toluene conversion over 94% and a H$_2$ production of 141.6 mol/mol toluene with no obvious deactivation observed in five-hour test was obtained.

Biofuels can be obtained through thermochemical routes based on the pyrolysis of biomass wastes and subsequent catalytic conditioning to reduce the oxygen content, a topic brilliantly addressed by García et al. Firstly, a two-step catalytic process (in-situ catalytic pyrolysis using CaO followed by a catalytic cracking of the vapours released using ZSM-5 zeolites) was used to directly obtain an upgraded bio-oil with an oxygen content of 16.4 wt.%. The resulting bio-oil was mixed with ethanol and gasoline in a 2/8/90 vol% ratio, and its behavior was studied in a stationary spark ignition engine, showing similar fuel consumption than pure gasoline at the same engine conditions. Additional benefits from gasoline blending with bio-oil and ethanol were a reduction in both the PAH and the carcinogenic equivalent concentrations, thus decreasing the environmental impact of the exhaust gases.

A different strategy for bio-oil upgrading based on the hydrodeoxygenation (HDO) reaction was used in the contribution of Suelves et al. In this work, a series of different carbon materials (namely, carbon nanofibers, carbon nanotubes, graphene oxide and activated carbons) were used as support for the preparation of Mo_2C based catalyst via carbothermal hydrogen reduction (CHR). The differences in the catalyst characteristics and their catalytic behavior were addressed using guaiacol, an aromatic model compound of fast pyrolysis bio-oils. Mo_2C catalyst supported on carbon nanofibers showed the best catalytic activity, with the highest selectivity to oxygen-free HDO products and guaiacol conversion as compared with other carbon-supported catalyst tested. This performance was related to its higher gasification resistance, good Mo_2C dispersion and crystal formation in CHR, which makes it the most suitable carbon material support for Mo_2C-based catalysts, paving the way for the deployment of HDO bio-oil upgrading technologies.

The use of novel catalyst to produce valuable chemicals from renewable resources is addressed in the three following contributions. Thus, Solsona et al studied the catalytic synthesis of γ-Valerolactone (GVL), a valuable chemical that can be used as a clean additive for automotive fuels, via dehydration and hydrogenation of levulinic acid, a biomass derived compound. A series of natural and low-cost clay-materials were used as Ni support. Different strategies aiming to avoid the addition of pressurized H_2 were pursued, such as the use of Zn in the aqueous media and the addition of an H_2 donor molecule (formic acid). Best results were obtained using Zn in the reaction media at 180° C and Ni supported on attapulgite or on a high surface area sepiolite as catalysts, leading to γ-valerolactone yields higher than 98%. The cellulose conversion into sorbitol via hydrolytic hydrogenation was addressed by Román–Martínez et al using mesoporous activated carbon supported Ru catalysts. Although a positive effect of a large amount of acidic oxygen surface groups was foreseen, such groups promoted the formation of by-products, and lower the sorbitol selectivity.

Therefore, highest sorbitol selectivity (91%) at 52% cellulose conversion was obtained under relatively mild reaction conditions with the pristine commercial mesoporous AC without further treatment. Another approach was used by Taylor et al to convert glycerol, a byproduct of biodiesel synthesis, into methanol and other useful chemicals, in a one-step low pressure process without the addition of hydrogen gas. A careful analysis of the process conditions was carried out and the influence of the surface area of CeO_2 as redox catalysts on their performance for glycerol conversion was addressed. The importance of the morphology of the catalyst and its impact on the of the reactivity of glycerol and its intermediates was proposed.

The last contribution of this special issue focused on the prospects of obtaining high quality hydrochar from red jujube branch, a by-product jujube industry, via hydrothermal carbonization. The hydrochar synthesis was optimized in terms of carbonization temperature and solid residence time, resulting in a material with very interesting properties as solid biofuel.

In conclusion, the papers collected in this Special Issue provide a very comprehensive snapshot about some of the most interesting trends in the catalysis area, such as the use of carbon supported catalyst, nickel as a replacement of noble metal and ceria as redox material. It is striking that most of the work here published are focused on the transformation of biomass derived molecules, clearly reflecting the enormous deal of attention that the technologies related to the development of the catalytic process in the context of biorefinery is gaining in the last years. This gives also a glimpse of the promising prospects of the catalysis in the area of biomass conversion technologies. We hope that all the reader enjoys these outstanding collections provided by our colleagues from some of the most prestigious research centers and universities wordwide, as we did as editing this special issue. Of course, all our gratitude for their outstanding work and implication. Special thanks to Ms. Vickie Zhang and Ms. Sally Xu for all their support during the edition of this special issue and its reprint as book format.

José Luis Pinilla, Isabel Suelves, Tomás García
Editors

Article

Biogas Upgrading Via Dry Reforming Over a Ni-Sn/CeO$_2$-Al$_2$O$_3$ Catalyst: Influence of the Biogas Source

Estelle le Saché, Sarah Johnson, Laura Pastor-Pérez, Bahman Amini Horri and Tomas R. Reina *

Chemical & Process Engineering Department, University of Surrey, Guildford GU2 7XH, UK; e.lesache@surrey.ac.uk (E.l.S.); sj00174@surrey.ac.uk (S.J.); l.pastorperez@surrey.ac.uk (L.P.-P.); b.aminihorri@surrey.ac.uk (B.A.H.)
* Correspondence: t.ramirezreina@surrey.ac.uk; Tel.: +44-148-368-6597

Received: 15 February 2019; Accepted: 13 March 2019; Published: 15 March 2019

Abstract: Biogas is a renewable, as well as abundant, fuel source which can be utilised in the production of heat and electricity as an alternative to fossil fuels. Biogas can additionally be upgraded via the dry reforming reactions into high value syngas. Nickel-based catalysts are well studied for this purpose but have shown little resilience to deactivation caused by carbon deposition. The use of bi-metallic formulations, as well as the introduction of promoters, are hence required to improve catalytic performance. In this study, the effect of varying compositions of model biogas (CH$_4$/CO$_2$ mixtures) on a promising multicomponent Ni-Sn/CeO$_2$-Al$_2$O$_3$ catalyst was investigated. For intermediate temperatures (650 °C), the catalyst displayed good levels of conversions in a surrogate sewage biogas (CH$_4$/CO$_2$ molar ratio of 1.5). Little deactivation was observed over a 20 h stability run, and greater coke resistance was achieved, related to a reference catalyst. Hence, this research confirms that biogas can suitably be used to generate H$_2$-rich syngas at intermediate temperatures provided a suitable catalyst is employed in the reaction.

Keywords: biogas; syngas production; DRM; Ni catalyst; bi-metallic catalyst; ceria-alumina

---

1. Introduction

Rising greenhouse gas (GHG) emissions and the associated global warming threat are some of the largest challenges facing the world today. The energy sector alone accounts for two-thirds of total greenhouse gas emissions, and around 80% of carbon dioxide emissions [1]. The majority of CO$_2$ emissions are produced from the combustion of fossil fuels for the generation of energy, and from other industrial processes. With global energy demand growing rapidly, the need for the decarbonisation of the energy industry has never been greater. Although CO$_2$ is the primary greenhouse gas emitted, the detrimental impact of releasing methane, with a global warming potential (GWP) of 25 [2], into the atmosphere should not be underestimated.

In recent years, more efforts have been made by national governments to implement GHG mitigating processes and technologies, such as carbon capture and storage. However, there is still much to be done to meet the target agreed on by 'The Paris Agreement' of limiting the global average temperature to below 2 °C above pre-industrial levels. Therefore, it is crucial to redirect the global focus from fossil fuels to renewable energy sources in order to minimise the destructive effects of climate change.

Biogas is one possible source of renewable energy and is commonly referred to as a gaseous mixture of methane and carbon dioxide produced by the anaerobic digestion of biodegradable matter. Typical feedstocks for biogas are waste materials, such as municipal waste, sewage sludge and agricultural waste. Other components present in biogas include water vapour, nitrogen and hydrogen sulphide. However, the actual composition, as well as the CH_4/CO_2 ratio, differs depending on the type of feedstock and the digestion process used [3]. Although biogas is the main renewable energy source contributing to the global energy supply [4], technical challenges still prevent it from being fully utilised for the generation of electricity.

Alternatively, biogas can be converted into high value syngas, a gaseous mixture of carbon monoxide and hydrogen, via the dry reforming of methane (DRM), as shown in Equation (1).

$$CH_4 + CO_2 \leftrightarrow 2CO + 2H_2 \quad \Delta H_{298K} = +247 \text{ kJ/mol} \tag{1}$$

Syngas is an extremely useful intermediate as it is used as a precursor to synthesise valuable fuels and chemicals such as methanol, as well as long chained hydrocarbons via the Fischer–Tropsch process [5,6]. Furthermore, syngas has been suggested as a feed gas to high temperature solid oxide fuel cells (SOFCs) with either internal or external reforming to generate electricity [7–9]. Figure 1 illustrates the opportunity to generate renewable energy from organic waste by combining anaerobic digestion with DRM and SOFCs. The high operating temperatures of SOFCs makes direct internal reforming feasible. However, the direct use of biogas in SOFCs is still associated with carbon deposition and sulphur poisoning in the fuel cell anode, which weakens cell durability and causes loss of cell performance [10].

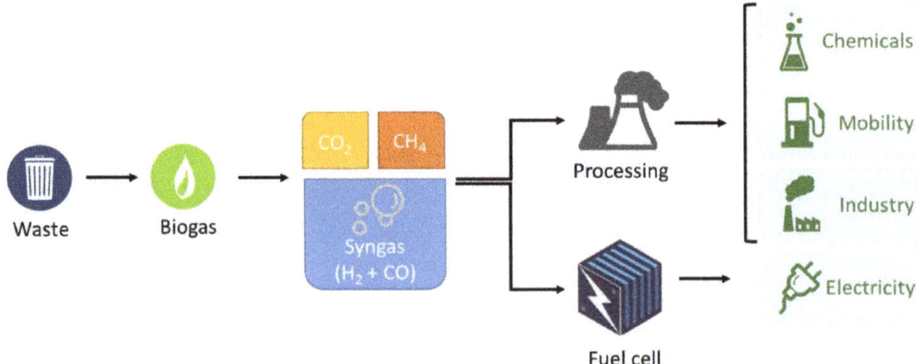

Figure 1. Block flow diagram illustrating the potential of utilising organic waste to generate 'green energy'.

Due to the endothermic nature of the dry reforming reaction and the high stability of the reactants, high reaction temperatures and a stable catalyst are required to achieve high syngas yields [11]. However, high reaction temperatures often lead to the deactivation of the catalyst. Deactivation is caused by either solid carbon deposition forming on the catalyst, which physically blocks the active metal phase and is known as coking, or through sintering of the catalyst active phase at high temperatures. There are a number of side reactions which are responsible for the formation of carbon, and hence negatively affect the performance of the catalyst. In dry reforming, carbon is most commonly produced by the Boudouard reaction, as shown in Equation (2), methane decomposition, as shown in Equation (3) as well as carbon monoxide reduction, as shown in Equation (4).

$$2CO \leftrightarrow C + CO_2 \quad \Delta H^{\circ}_{298} = -171 \text{ kJ mol}^{-1} \tag{2}$$

$$CH_4 \leftrightarrow C + 2H_2 \quad \Delta H^{\circ}_{298} = 75 \text{ kJ mol}^{-1} \tag{3}$$

$$CO + H_2 \leftrightarrow C + H_2O \quad \Delta H^{\circ}_{298} = -131 \text{ kJ mol}^{-1} \tag{4}$$

Therefore, an effective dry reforming catalyst needs to be thermally stable in order to be resistant to coking and sintering, whilst yielding optimal conversions. Additionally, it must be economically viable so that it can be cost effectively scaled-up for industrial applications. All these factors are heavily dependent on the selected active metal, as well as the properties of the support and/or promoter [5].

Nickel-based materials are widely regarded as effective catalysts for the dry reforming of methane as they display good activity and respectable conversions [12]. However, the main drawback of Ni-based catalysts is that they suffer from quick deactivation as a result of carbon deposition and an inclination to sintering. Noble metal-based catalysts are frequently cited as a superior alternative due to greater stability and higher resilience to coking. Catalysts based on Ru and Rh in particular show higher activity compared to nickel [13]. Despite this, noble metal catalysts are far more expensive and have limited availability in comparison to Ni-based ones and are therefore unfeasible for use in industrial applications.

In order to develop a stable, high performing nickel-based catalyst, the material is often upgraded using an appropriate support or adding a promoter. Alumina (Al_2O_3)-based supports have been thoroughly investigated due to their high specific area which improves metal dispersion [14]. Yet implementing alumina as a support is still associated with catalyst deactivation, caused mainly by coke formation on the acid sites [14]. To combat this problem, basic promoters such as ceria (CeO_2) can be used to stabilise the support. It has been shown that Ni catalysts supported on CeO_2-Al_2O_3 systems showed far better performance than either CeO_2 or Al_2O_3 supported nickel catalysts [15]. Indeed, ceria not only tunes the acid/base properties of the support but also provides excellent oxygen mobility (due to its high oxygen storage capacity), thus helping to prevent coking via oxidation of the coke precursors [16].

Moreover, bi-metallic systems can further enhance catalytic performance; the addition of a second metal has shown to improve the activity, as well as the stability of the catalyst [17–19]. It is believed that this is related to the metal–metal interactions which reduce the electron donor capacity of the active metal, limiting its tendency to form strong bonds with carbon precursors [20,21]. As recently demonstrated by Stroud et al., bi-metallic Ni-Sn supported Al_2O_3 and CeO_2-Al_2O_3 catalysts can effectively generate good conversions for DRM whilst exhibiting high stability [16]. The positive effect of Sn as an added metal to Ni was further investigated by Guharoy et al. in a DFT study concluding that the beneficial effect of tin is credited to Sn atoms occupying C nucleation sites in the vicinity on Ni atoms, slowing coke formation (i.e., increasing the energy barrier for coke nucleation) [22].

Under these premises, the aim of this work is to investigate the performance of an optimised multicomponent Ni-Sn/CeO_2-Al_2O_3 catalyst for the production of syngas from a biogas feed via the dry reforming of methane. The impact of the type of biogas and more precisely its origin (i.e., sewage, municipal waste, landfill and organic waste) is also a subject of this study which aims to showcase a suitable upgrading route for different types of biogases via reforming reactions.

2. Materials and Methods

2.1. Catalyst Preparation

The Ni-Sn/CeO_2-Al_2O_3 catalyst was synthesised by sequential impregnation. First, the ceria promoted support was prepared by impregnation of the correct amount of $Ce(NO_3)_2 \cdot 6H_2O$ (Sigma-Aldrich) on γ-Al_2O_3 powder (Sasol) in order to obtain a 20 wt % loading of CeO_2. After 1 h of stirring, the solvent, acetone, was evaporated at reduced pressure in a rotary evaporator. The resulting powder was dried overnight at 80 °C and calcined at 800 °C for 4 h. The support was then impregnated in the same way with the necessary amount of $Ni(NO_3)_2 \cdot 6H_2O$ (Sigma-Aldrich) dissolved in acetone, to achieve a 10 wt % metal loading. Lastly, the obtained solid was impregnated with $SnCl_2$ (Sigma-Aldrich) following the same procedure to achieve a Sn/Ni molar ratio of 0.02. This ratio was chosen based on previous work [16]. For simplicity the catalyst will be referred to as Ni-Sn/CeAl from here on.

2.2. Catalyst Characterisation

N_2-adsorption-desorption analysis was performed in a QuadraSorb Station 4 at liquid nitrogen temperature. Prior to the analysis, the catalyst was degassed at 250 °C for 4 h in vacuum. The surface area was calculated from the Brunauer–Emmett–Teller (BET) equation.

X-ray fluorescence (XRF) analysis of the catalyst was carried out on an EDAX Eagle III spectrophotometer using rhodium as the radiation source.

X-ray diffraction (XRD) analysis was conducted on fresh and spent samples using an X'Pert Powder instrument from PANalytical. The diffraction patterns were recorded over a 2θ range of 10 to 90°. A step size of 0.05° was used with a time step of 450 s. Diffraction patterns were recorded at 30 mA and 40 kV, using Cu Kα radiation (λ = 0.154 nm).

Temperature-programmed reduction (TPR) in hydrogen was carried out on the calcined sample in a U-shaped quartz cell using a 5% H_2/He gas flow of 50 mL·min^{-1}, with a heating rate of 10 °C min^{-1}. Prior to analysis the catalyst was treated with He at 150 °C for 1 h. Hydrogen consumption was measured by on-line mass spectrometry (Pfeiffer, OmniStar GSD 301).

Thermogravimetric analysis (DSC-TGA) was carried out on the samples post stability test in an SDT Q600 V8 from TA Instruments. The samples were ramped from room temperature to 900 °C at 10 °C min^{-1} in air.

2.3. Catalytic Activity

The catalytic performance for the dry reforming of methane at varying compositions of model biogas (CH_4 and CO_2) was carried out in a continuous flow quartz tube reactor (10 mm ID) equipped with a thermocouple, at atmospheric pressure. Reactants and products were followed by an on-line gas analyser (ABB AO2020). In order to achieve a weight hourly space velocity (WHSVs) of 30 L g^{-1} h^{-1}, a mass of 0.2 g of catalyst was used for each screening. All samples were pre-reduced in situ using 100 mL·min^{-1} of 10% H_2 in N_2 at 850 °C for 1 h.

Reactions were conducted using a total flow of 100 mL·min^{-1} of varying biogas compositions for temperatures ranging from 550 to 850 °C in 50 °C increments. The catalytic performance was measured for CH_4/CO_2 molar ratios of 1, 1.25, 1.5 and 1.85 balanced in N_2. These ratios were carefully chosen in order to model DRM of biogas produced from a range of residues [23]. Table 1 lists the types of biogas source, as well as the corresponding CH_4 and CO_2 content, being modelled by the specific CH_4/CO_2 molar ratio selected for each reaction.

Table 1. List of biogas sources and the corresponding methane and carbon dioxide content, and the resulting CH_4/CO_2 molar ratio. Adapted from Lau et al. [23].

Biogas Source	CH$_4$ Content (%)	CO$_2$ Content (%)	CH$_4$/CO$_2$ Molar Ratio
Model Composition	50	50	1:1
Landfill Waste	55	45	1.25
Sewage Waste	60	40	1.5
Organic Waste	65	35	1.85

Lastly, stability tests were carried out on Ni-Sn/CeAl and a reference nickel on alumina with the same Ni loading (Ni/Al) catalyst using CH_4/CO_2 molar ratio of 1 and 1.5 and a temperature of 650 °C for 20 h. The low temperature was chosen in order to simulate a syngas suitable for direct use in intermediate temperature SOFCs aiming for the hypothetical utilisation of biogas in an internal reforming SOFC.

Conversions of the reactants, as well as the H_2/CO ratio and H_2 yields, were calculated for each run to determine the effect of varying the biogas composition on the performance of the catalyst as follows:

$$X_{CH4}(\%) = 100 * \frac{[CH_4]_{In} - [CH_4]_{Out}}{[CH_4]_{In}} \tag{5}$$

$$X_{CO2}(\%) = 100 * \frac{[CO_2]_{In} - [CO_2]_{Out}}{[CO_2]_{Out}} \tag{6}$$

$$H_2/COratio = \frac{[H_2]_{Out}}{[CO]_{Out}} \tag{7}$$

$$Y_{H2}(\%) = 100 * \frac{[H_2]_{Out}}{2[CH_4]_{In}} \tag{8}$$

3. Results and Discussion

3.1. Physicochemical Properties

The N_2 adsorption-desorption isotherm of the calcined catalyst is shown in Figure 2. The analysis generated a type IV isotherm with a characteristic hysteresis loop, which is associated with the presence of well-developed cylindrical mesopores. The steepness of the loop suggests that the mesopores were homogeneously distributed throughout the structure of the sample [24].

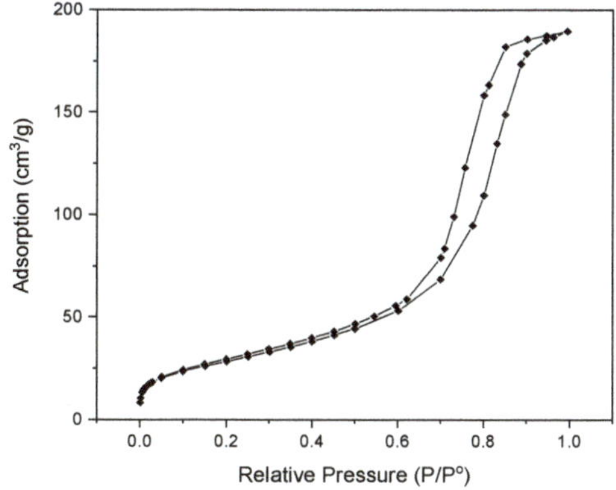

Figure 2. Nitrogen adsorption-desorption isotherm of Ni-Sn/CeAl sample.

The surface area, calculated using the BET method, the total pore volume and the average pore diameter of the calcined catalyst are listed in Table 2 along with its elemental composition determined by XRF. Although the textural properties were primarily governed by the γ-Al_2O_3 support, the surface area and pore volume measured were smaller than that of the support itself due to CeO_2 and Ni particles present on the surface and covering the pores [25]. The metal content of the obtained catalyst was very close to the intended value.

Table 2. Textural properties of Ni-Sn/CeAl post calcination as determined by the Brunauer–Emmett–Teller (BET) equation and X-ray fluorescence (XRF) analysis.

	NiO wt %	Sn wt %	S_{Bet} (m²·g⁻¹)	V_{Pore} (cm³·g⁻¹)	D_{Pore} (nm)
Ni-Sn/CeAl	10.3	1.3	103	0.31	8.3
CeAl	-	-	130	0.34	9.6

3.2. Reducibility: H_2-TPR

The temperature programmed reduction was conducted in order investigate the interactions between the support and metallic species, as well as the redox properties of the catalyst which are essential for DRM [5]. Figure 3 shows the H_2-TPR profile of the calcined catalyst. The Ni-Sn/CeAl sample presents a main reduction peak at around 820 °C. This high reduction temperature is attributed to the reduction of Ni^{2+} to Ni, strongly interacting with the alumina support. The high reduction temperature may indicate the presence of $NiAl_2O_4$ spinel, which is difficult to reduce and is associated with high reduction temperatures of 600–870 °C [16,26]. It appears that a second reduction process occurs at 900 °C, possibly due to the reduction of CeO_2 and Al_2O_3 to $CeAlO_3$, in good agreement with the XRD results that will be discussed later in the paper [27,28].

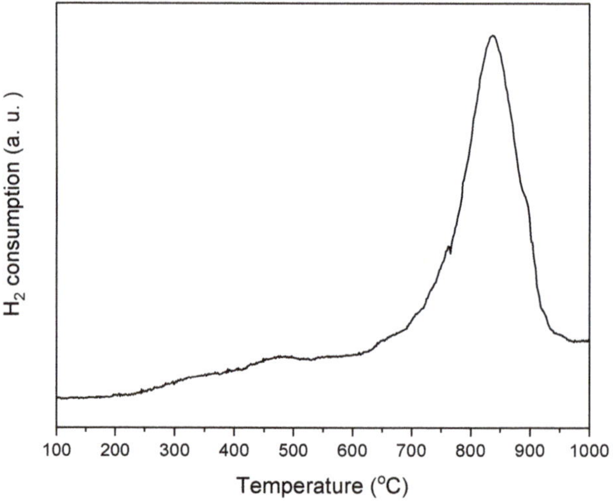

Figure 3. H_2-TPR profile of Ni-Sn/CeAl.

3.3. XRD

The structural properties of the Ni-Sn/CeAl catalyst after calcination (fresh) and after reduction under H_2 at 850 °C for 1 hour were determined and the XRD profiles are shown in Figure 4. The fresh catalyst profile shows no presence of characteristic crystalline peaks associated to Sn or SnO_x due to the low amount used, metallic nickel (Ni^0) or oxidised nickel (NiO) species. The profile presents, however, peaks related to γ-Al_2O_3 (JCPDS# 00-048-0367) and $NiAl_2O_4$ (JCPDS# 00-010-0339). Although γ-Al_2O_3 and $NiAl_2O_4$ phases are hardly distinguishable due to their broad and overlapping diffraction peaks, the formation of nickel aluminate spinel ($NiAl_2O_4$) can be identified due to the slight shifts towards smaller angles of three strong diffraction peaks associated with γ-Al_2O_3: at 2θ 37.0°, 45.5° and 66.3° [27]. Indeed, the nickel aluminate spinel can be formed at high calcination temperature (800 °C) via the reaction between NiO and Al_2O_3. Additionally, for the full formation of aluminate spinel, a Ni loading of around 33 wt % is necessary [29], therefore, since the Ni loading in this sample was much lower it is most likely that $NiAl_2O_4$ spinel co-exists with Ni nanoparticles supported on the γ-Al_2O_3 support.

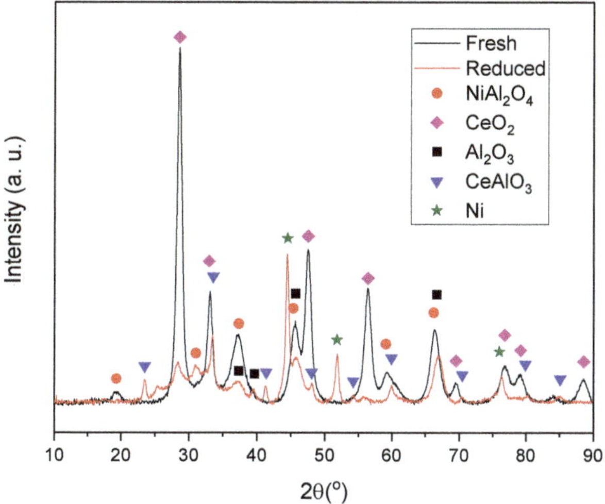

Figure 4. XRD profiles of the fresh and reduced Ni-Sn/CeAl catalyst.

The fresh sample presents, as well, the typical diffraction peaks of CeO_2 fluorite cubic cells (JCPDS# 01-075-0390) at 28.5°, 33.1°, 47.8° and 56.3°. However, once the sample was reduced, an additional $CeAlO_3$ tetragonal phase (JCPDS# 01-081-1185) was detected. When reduced at temperatures above 600 °C, Ce_2O_3 and γ-Al_2O_3 reacted to produce $CeAlO_3$ in good agreement with the results discussed in the TPR section [28]. However not all CeO_x species were reacted to $CeAlO_3$ since traces of CeO_2 cubic phase can still be detected at 28.5°. Finally, the XRD pattern of the sample reduced in hydrogen shows the presence of metallic Ni particles (JCPDS# 87-0712), which indicates that Ni^0 will be the predominant active phase for the dry reforming reaction. The shift towards the higher angles of the 37.0°, 45.5° and 66.3° peaks confirms that $NiAl_2O_4$ has been reduced to Ni metallic and γ-Al_2O_3 upon reduction at 850 °C, confirming the TPR results.

3.4. Catalyst Performance

The catalytic performance for DRM was studied over a range of temperatures (550–850 °C) at CH_4/CO_2 molar ratios of 1, 1.25, 1.5 and 1.85. The effect of varying the biogas feed concentration, in terms of CH_4 conversion as a function of temperature, is reported in Figure 5a. It was shown that with increasing methane concentration, the overall CH_4 conversion decreases. Indeed, methane was introduced in excess for DRM. Additionally, CH_4 conversion increased with temperature; this aligns with expectations, as the endothermic nature of the reaction requires higher temperatures to reach equilibrium conversion. Hence, superior activity of the catalyst was observed during the temperature screening at a 1:1 molar ratio of CH_4/CO_2, which reached a methane conversion of 93% at 850 °C.

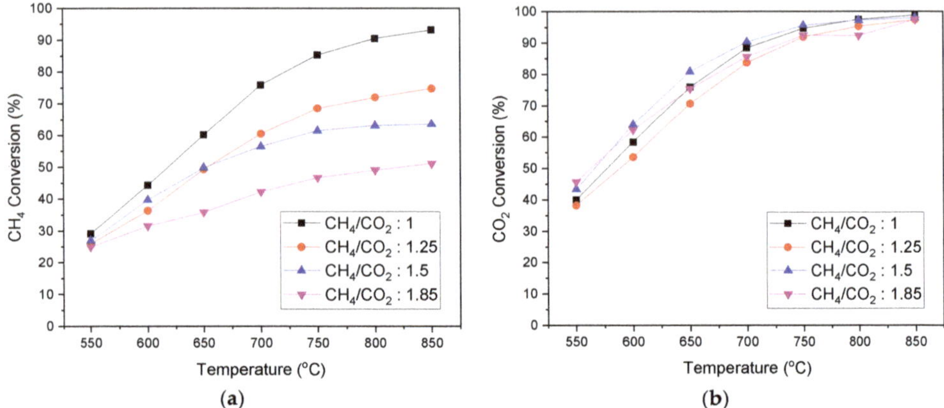

Figure 5. Catalytic performance of Ni-Sn/CeAl at CH_4/CO_2 molar ratios of 1, 1.25, 1.5 and 1.85 as a function of temperature: (**a**) methane conversion; (**b**) carbon dioxide conversion.

Figure 5b shows the conversion of CO_2 as a function of temperature for various biogas compositions. The temperature screening carried out at CH_4/CO_2 molar ratio of 1.5 displayed the highest conversion for most temperatures. For all ratios, the highest CO_2 conversion was achieved at 850 °C, reaching around 98%. Overall, the conversion witnessed for CO_2 was much higher than that of CH_4; CO_2 being the limiting reactant for ratios greater than 1. For the model biogas mixture (CH_4/CO_2 = 1) however, CH_4 and CO_2 conversions should be similar. The fact that CO_2 conversion was slightly higher than CH_4 conversion for temperatures lower than 850 °C was due to the occurrence of the reverse water gas shift (RWGS) reaction, consuming CO_2 and H_2 to form CO as previously reported elsewhere [5]. This can also be attributed to the high activation energy of methane [11,22].

The effect of varying the biogas feed concentration on the H_2/CO ratio is illustrated in Figure 6a. As a general rule, the H_2/CO ratio increased with higher temperatures, with the exception of the temperature screening carried out at a CH_4/CO_2 molar ratio of 1.85, where the H_2/CO ratio initially decreased with the increment of the temperature and only increased once a temperature of 650 °C was reached. The high concentration of hydrogen in the products stream may be explained by the excess of methane in the feed gas. It is likely that this excess created favourable conditions for the methane decomposition reaction to take place, which contributed to a higher concentration (partial pressure) of H_2 at lower temperatures. Furthermore, it was observed that at lower temperatures (550–600 °C), a larger CH_4/CO_2 ratio yielded a higher H_2/CO ratio. Whereas, at higher temperatures (700–850 °C), the H_2/CO ratio exhibited the largest value for a model biogas feed of 1:1 molar ratio of CH_4/CO_2.

The H_2 yield obtained for each biogas composition as a function of temperature is shown in Figure 6b. As expected, yields increased along with temperature and CH_4 conversion. As the CH_4/CO_2 molar ratio increased, H_2 yields decreased due to excess CH_4 in the feed and reduced CH_4 conversion.

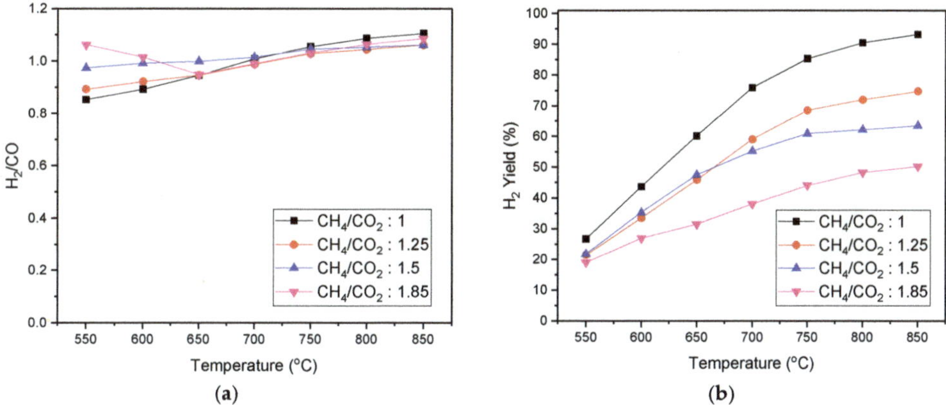

Figure 6. (a) H_2/CO ratio and (b) H_2 yield for Ni-Sn/CeAl at CH_4/CO_2 molar ratios of 1, 1.25, 1.5 and 1.85 as a function of temperature.

In order to establish a more general comparison with previously reported Ni-based catalysts for biogas reforming reactions, different reported catalysts and the conditions used in the biogas dry reforming are presented in Table 3. As shown in the table, our Ni-Sn/CeAl catalyst can be deemed to perform with superior behaviour in the compared conditions, especially in terms of hydrogen yields. Furthermore, our catalyst continues showing a good performance tested at lower temperatures and long-term tests, which reinforces the exceptional behaviour of the developed system.

Table 3. Catalysts employed, and conditions used in catalytic biogas dry reforming, using CH_4/CO_2 ratio: 1.5. WHSV: weight hourly space velocity.

	Temperature (°C)	WHSV ($L \cdot g^{-1} h^{-1}$)	CH_4 Conversion (%)	CO_2 Conversion (%)	H_2 Yield (%)	Reference
Ni/MgO	800	60	62.2	92.3	52.6	[30]
Ni/Ce-Al$_2$O$_3$	800	60	59.3	77.8	57.1	[30]
Ni/Zr-Al$_2$O$_3$	800	60	58.4	73.5	51.9	[30]
Ni/Ce-Zr-Al$_2$O$_3$	800	60	64.5	85.6	57.5	[30]
Rh-Ni/Ce-Al$_2$O$_3$	800	60	60.1	94.4	63.3	[30]
Ni/Al$_2$O$_3$	800	120	60.0	83.0	48.0	[14]
Ni/Ce-Al$_2$O$_3$	800	120	63.0	89.0	50.0	[14]
Ni/La-Al$_2$O$_3$	800	120	69.0	94.0	57.0	[14]
Ni/Ce-La-Al$_2$O$_3$	800	120	69.0	94.5	58.0	[14]
Ni/CaO-Al$_2$O$_3$	800	120	57.9	75.2	30.5	[6]
Ni/MgO-Al$_2$O$_3$	800	120	57.7	75.2	32.4	[6]
Reformax® 250	750	18	64.0	86.0	28.8	[31]
Ni-Sn/Ce–Al$_2$O$_3$	800	30	63.1	97.5	62.3	This work
Equilibrium	800	-	65.5	99.1	65.2	

3.5. Stability Test

Long-term stability tests were carried out at a temperature of 650 °C. The temperature was selected in order to simulate a syngas suitable for direct use i+n intermediate temperature SOFCs.

The results of the stability tests are displayed in Figure 7, which shows both CH_4 and CO_2 conversion as a function of time. Figure 7a compares the stability of Ni-Sn/CeAl under two biogas mixtures: 1 and 1.5 CH_4/CO_2 molar ratios. When tested with the model biogas mixture, the catalyst showed good performance with 79% CO_2 conversion and 60% CH_4 conversion. The performance was stable with a slight deactivation equivalent to 5% conversion loss. On the other hand, when the feed CH_4/CO_2 molar ratio was increased to 1.5, CH_4 conversion was much lower, as previously observed, but the deactivation of the catalyst, although not drastic, was more pronounced than the one of the model mixture. The excess of methane in the feed stream seems to promote methane decomposition and therefore enhance the coking of the catalyst. This resulted in the loss of 10% and 15% conversion for CO_2 and CH_4, respectively.

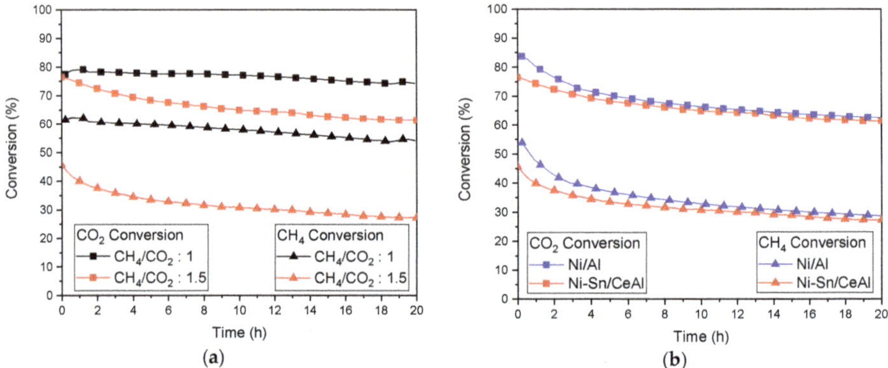

Figure 7. Dry reforming of methane (DRM) stability test at 650 °C on (**a**) Ni-Sn/CeAl at CH_4/CO_2 molar ratios of 1 (model biogas) and 1.5 (sewage waste) and (**b**) Ni-Sn/CeAl and Ni/Al catalyst using CH_4/CO_2 molar ratios of 1.5: CO_2 and CH_4 conversions.

The long-term test using sewage waste biogas was compared to a reference Ni/Al_2O_3 catalyst on Figure 7b. The reference catalyst seems to perform slightly better than the promoted catalyst with slightly higher conversions. In fact, the introduction of Sn in the formulation of the catalyst was meant to prevent coke formation since Sn has a similar electronic configuration and was nucleating with Ni. Nevertheless, such an interaction may reduce the catalytic activity of the catalyst. Indeed, the deactivation of the reference catalysts was clearly more pronounced than that observed in the multicomponent catalyst, again highlighting the robustness of the Ni-Sn catalysts for long runs.

3.6. Study of Carbonaceous Deposits

Thermogravimetric analysis was conducted on the samples after reacting for 20 h to estimate the coke formation during reaction. Figure 8a shows the effect of biogas mixture on coke deposition. When subjected to the model biogas mixture, around 0.256 g_C/g_{cat} was formed whereas for sewage derived biogas, 0.268 g_C/g_{cat} was formed. The catalyst displayed faster deactivation under methane rich biogas, confirming that methane decomposition, as shown in Equation (3), must have been favoured, hence forming more coke on the catalyst. Figure 8b compares the carbon loading on two different catalysts after reacting with methane-rich biogas. The reference catalyst Ni/Al had similar catalytic performance to the Ni-Sn/CeAl catalyst, however, it was more prone to coking with 0.346 g_C/g_{cat}. The multicomponent catalyst benefits from the high oxygen storage capacity of ceria that facilitates carbon oxidation [5]. In addition, Sn atoms were occupying C nucleation sites in the vicinity of Ni, preventing carbon from poisoning the active sites [16]. Moreover, the gasification of the coke present on the Ni/Al catalyst occurred at higher temperatures, suggesting the formation of more graphitic carbon as previously observed in the literature [16].

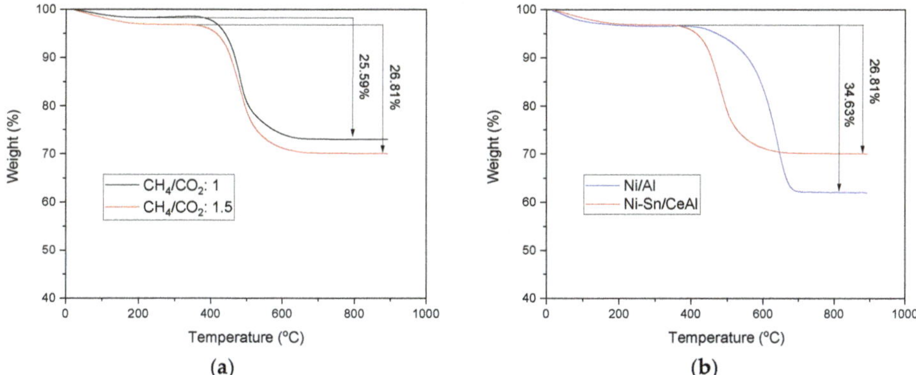

Figure 8. Thermogravimetric analysis for the catalysts after (**a**) Ni-Sn/CeAl at CH_4/CO_2 molar ratios of 1 (model biogas) and 1.5 (sewage waste) and (**b**) Ni-Sn/CeAl and Ni/Al catalyst using CH_4/CO_2 molar ratios of 1.5.

3.7. Post Reaction Characterisation

Figure 9 shows the XRD profiles of the Ni-Sn/CeAl catalyst after reduction, after reacting from 550 to 850 °C and after the 20 h stability run at 650 °C in a biogas ratio of $CH_4/CO_2 = 1.5$. The reduced sample, as described earlier, presents the characteristic peaks of $CeAlO_3$, Ni, CeO_2 and Al_2O_3. After reacting at temperatures up to 850 °C, CeO_2 seems to have completely reacted with Al_2O_3 to form $CeAlO_3$ exclusively. The high temperatures and reducing atmosphere associated with the reaction conditions favoured the phase transition. After the stability test however, the CeO_2 cubic phase displayed peaks of higher intensities than those of the reduced catalyst and the post temperature screening reaction catalyst. The low temperature of the stability test, 650 °C, seems to have favoured the reverse reaction and allowed $CeAlO_3$ to re-oxidise back to CeO_2. Indeed, the peaks associated with $CeAlO_3$ decreased in intensity while the CeO_2 diffraction peaks increased. Additionally, all post mortem samples presented a large peak at 26° associated to graphitic carbon. In terms of Ni crystallite size, Ni was estimated using the Scherrer equation at 17 nm, 20 nm and 11 nm in the sample reduced, after the temperature-dependant run and after the stability test, respectively. The temperature-dependant experiment was performed up to 850 °C, temperatures far above the Tammann temperature of Ni. Nickel sintering was prone to happen at such temperatures, explaining the increase in crystallite size. However, a reduction in the nickel crystallite size was observed after the 20 h stability test. First, the test took place at a much lower temperature (650 °C), which may have slowed down sintering. Second, the reaction atmosphere has an influence on the interactions amongst Ni, CeO_2 and Al_2O_3. The re-oxidation of $CeAlO_3$ to form CeO_2 may have re-dispersed Ni crystallites on the catalyst surface. Zou et al. observed the same phenomenon on a Ni/ CeO_2-Al_2O_3 catalyst reduced at 800 °C and after 105 h reaction at 350 °C. It appears Ni crystallites were re-constructed on the catalyst surface under the reaction atmosphere [27].

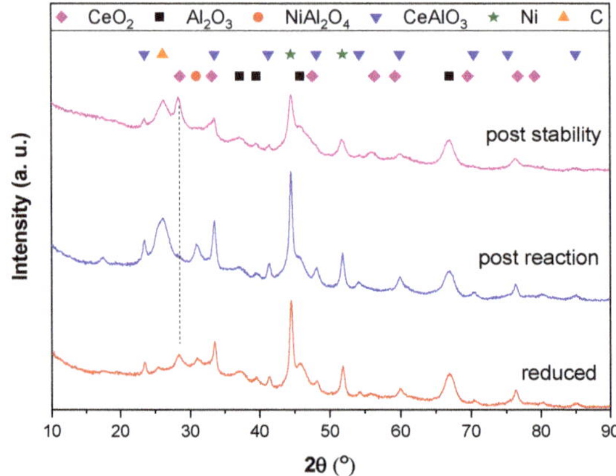

Figure 9. XRD characterisation spectra for Ni-Sn/CeAl reduced, post reaction and post stability at a CH$_4$/CO$_2$ molar ratio of 1.5.

4. Conclusions

In this work, a high performing Ni-Sn/CeO$_2$-Al$_2$O$_3$ catalyst was developed for the conversion of biogas to syngas via the dry reforming of methane. The synthesised material was based on the DRM catalytic standard of Ni/Al$_2$O$_3$, which is an inexpensive alternative to high performing noble metal-based catalysts, and upgraded using CeO$_2$ and Sn.

The performance of the multicomponent catalyst was investigated for a range of temperatures and model biogas compositions by determining the conversion of each reactant, as well as the H$_2$/CO ratio of the syngas produced and the H$_2$ yield. Additionally, multiple characterisation techniques were carried out on fresh, reduced and spent samples of catalyst, exhibiting the crystal structure changes and the reversible reducibility for the support.

Overall, the Ni-Sn/CeAl catalyst exhibited respectable conversions of both CO$_2$ and CH$_4$ ratios for all compositions of biogas. The effect of biogas composition on the stability of the catalyst was investigated, and although no extensive signs of deactivation were detected during 20 h, the presence of excess methane accelerated the deactivation of the catalyst due to additional methane decomposition. The stability of the catalyst was also compared with the standard Ni/Al$_2$O$_3$. Although both catalysts displayed similar catalytic activity, the multicomponent catalyst benefited from greater coke resistance, confirming the promotion effect of both tin and ceria.

In summary, our multicomponent catalyst is a suitable material—highly active and considerably stable—to be implemented in a biogas upgrading unit to generate renewable energy (i.e., using a SOFC) or added-value products using syngas as a platform chemical.

Author Contributions: Conceptualization, T.R.R. & B.A.H.; methodology, L.P.-P.; formal analysis, E.l.S.; investigation, S.J. & E.l.S.; writing—original draft preparation, S.J.; writing—review and editing, E.l.S. & T.R.R.; supervision, T.R.R. & B.A.H.; funding acquisition, T.R.R.

Funding: This research was funded by the Department of Chemical and Process Engineering at the University of Surrey and by the EPSRC, grant EP/R512904/1 as well as the Royal Society, Research Grant RSGR1180353. LPP acknowledge Comunitat Valenciana for her APOSTD2017 fellowship. This work was also partially sponsored by the CO2Chem through the EPSRC grant EP/P026435/1.

Acknowledgments: The authors acknowledge Sasol for supplying the alumina support.

Conflicts of Interest: The authors declare no conflict of interest. The funders had no role in the design of the study; in the collection, analyses, or interpretation of data; in the writing of the manuscript, or in the decision to publish the results.

References

1. International Energy Agency. *CO$_2$ Emissions from Fuel Combustion–Highlights*; International Energy Agency: Paris, France, 2018.
2. Boucher, O.; Friedlingstein, P.; Collins, B.; Shine, K.P. The indirect global warming potential and global temperature change potential due to methane oxidation. *Environ. Res. Lett.* **2009**, *4*, 044007. [CrossRef]
3. Ahmed, S.; Lee, S.H.D.; Ferrandon, M.S. Catalytic steam reforming of biogas–effects of feed composition and operating conditions. *Int. J. Hydrogen Energy* **2015**, *40*, 1005–1015. [CrossRef]
4. Mota, N.; Alvarez-Galvan, C.; Navarro, R.; Fierro, J. Biogas as a source of renewable syngas production: Advances and challenges. *Biofuels* **2011**, *2*, 325–343. [CrossRef]
5. Le Saché, E.; Santos, J.L.; Smith, T.J.; Centeno, M.A.; Arellano-Garcia, H.; Odriozola, J.A.; Reina, T.R. Multicomponent Ni-CeO$_2$ nanocatalysts for syngas production from CO$_2$/CH$_4$ mixtures. *J. CO$_2$ Util.* **2018**, *25*, 68–78. [CrossRef]
6. Charisiou, N.D.; Baklavaridis, A.; Papadakis, V.G.; Goula, M.A. Synthesis gas production via the biogas reforming reaction over Ni/MgO–Al$_2$O$_3$ and Ni/CaO–Al$_2$O$_3$ catalysts. *Waste Biomass Valoriz.* **2016**, *7*, 725–736. [CrossRef]
7. Bonura, G.; Cannilla, C.; Frusteri, F. Ceria–gadolinia supported nicu catalyst: A suitable system for dry reforming of biogas to feed a solid oxide fuel cell (SOFC). *Appl. Catal. B* **2012**, *121–122*, 135–147. [CrossRef]
8. Lanzini, A.; Leone, P. Experimental investigation of direct internal reforming of biogas in solid oxide fuel cells. *Int. J. Hydrogen Energy* **2010**, *35*, 2463–2476. [CrossRef]
9. Shiratori, Y.; Ijichi, T.; Oshima, T.; Sasaki, K. Internal reforming sofc running on biogas. *Int. J. Hydrogen Energy* **2010**, *35*, 7905–7912. [CrossRef]
10. Fuerte, A.; Valenzuela, R.X.; Escudero, M.J.; Daza, L. Study of a sofc with a bimetallic cu–co–ceria anode directly fuelled with simulated biogas mixtures. *Int. J. Hydrogen Energy* **2014**, *39*, 4060–4066. [CrossRef]
11. Le Saché, E.; Pastor-Pérez, L.; Watson, D.; Sepúlveda-Escribano, A.; Reina, T.R. Ni stabilised on inorganic complex structures: Superior catalysts for chemical co$_2$ recycling via dry reforming of methane. *Appl. Catal. B* **2018**, *236*, 458–465. [CrossRef]
12. Pakhare, D.; Spivey, J. A review of dry (CO$_2$) reforming of methane over noble metal catalysts. *Chem. Soc. Rev.* **2014**, *43*, 7813–7837. [CrossRef]
13. Shah, Y.T.; Gardner, T.H. Dry reforming of hydrocarbon feedstocks. *Catal. Rev. Sci. Eng.* **2014**, *56*, 476–536. [CrossRef]
14. Charisiou, N.D.; Siakavelas, G.; Papageridis, K.N.; Baklavaridis, A.; Tzounis, L.; Avraam, D.G.; Goula, M.A. Syngas production via the biogas dry reforming reaction over nickel supported on modified with CeO$_2$ and/or La$_2$O$_3$ alumina catalysts. *J. Nat. Gas Sci. Eng.* **2016**, *31*, 164–183. [CrossRef]
15. Wang, S.; Lu, G.Q. Role of CeO$_2$ in Ni/CeO$_2$–Al$_2$O$_3$ catalysts for carbon dioxide reforming of methane. *Appl. Catal. B* **1998**, *19*, 267–277. [CrossRef]
16. Stroud, T.; Smith, T.J.; le Saché, E.; Santos, J.L.; Centeno, M.A.; Arellano-Garcia, H.; Odriozola, J.A.; Reina, T.R. Chemical CO$_2$ recycling via dry and bi reforming of methane using Ni-Sn/Al$_2$O$_3$ and Ni-Sn/CeO$_2$-Al$_2$O$_3$ catalysts. *Appl. Catal. B* **2018**, *224*, 125–135. [CrossRef]
17. San-José-Alonso, D.; Juan-Juan, J.; Illán-Gómez, M.J.; Román-Martínez, M.C. Ni, CO and bimetallic Ni–CO catalysts for the dry reforming of methane. *Appl. Catal. A* **2009**, *371*, 54–59. [CrossRef]
18. Zhang, J.; Wang, H.; Dalai, A.K. Development of stable bimetallic catalysts for carbon dioxide reforming of methane. *J. Catal.* **2007**, *249*, 300–310. [CrossRef]
19. Crisafulli, C.; Scirè, S.; Minicò, S.; Solarino, L. Ni–Ru bimetallic catalysts for the CO$_2$ reforming of methane. *Appl. Catal. A* **2002**, *225*, 1–9. [CrossRef]
20. Crisafulli, C.; Scirè, S.; Maggiore, R.; Minicò, S.; Galvagno, S. CO$_2$ reforming of methane over Ni–Ru and Ni–Pd bimetallic catalysts. *Catal. Lett.* **1999**, *59*, 21–26. [CrossRef]
21. Boldrin, P.; Ruiz-Trejo, E.; Mermelstein, J.; Bermúdez Menéndez, J.M.; Ramírez Reina, T.; Brandon, N.P. Strategies for carbon and sulfur tolerant solid oxide fuel cell materials, incorporating lessons from heterogeneous catalysis. *Chem. Rev.* **2016**, *116*, 13633–13684. [CrossRef]
22. Guharoy, U.; Le Saché, E.; Cai, Q.; Reina, T.R.; Gu, S. Understanding the role of Ni-Sn interaction to design highly effective CO$_2$ conversion catalysts for dry reforming of methane. *J. CO$_2$ Util.* **2018**, *27*, 1–10. [CrossRef]

23. Lau, C.S.; Tsolakis, A.; Wyszynski, M.L. Biogas upgrade to syn-gas (H_2–CO) via dry and oxidative reforming. *Int. J. Hydrogen Energy* **2011**, *36*, 397–404. [CrossRef]

24. Tao, K.; Shi, L.; Ma, Q.; Wang, D.; Zeng, C.; Kong, C.; Wu, M.; Chen, L.; Zhou, S.; Hu, Y.; et al. Methane reforming with carbon dioxide over mesoporous nickel–alumina composite catalyst. *Chem. Eng. J.* **2013**, *221*, 25–31. [CrossRef]

25. Yang, L.; Pastor-Pérez, L.; Gu, S.; Sepúlveda-Escribano, A.; Reina, T.R. Highly efficient Ni/CeO$_2$-Al$_2$O$_3$ catalysts for CO_2 upgrading via reverse water-gas shift: Effect of selected transition metal promoters. *Appl. Catal. B* **2018**, *232*, 464–471. [CrossRef]

26. Li, P.-P.; Lang, W.-Z.; Xia, K.; Yan, X.; Guo, Y.-J. The promotion effects of Ni on the properties of Cr/Al catalysts for propane dehydrogenation reaction. *Appl. Catal. A* **2016**, *522*, 172–179. [CrossRef]

27. Zou, X.; Wang, X.; Li, L.; Shen, K.; Lu, X.; Ding, W. Development of highly effective supported nickel catalysts for pre-reforming of liquefied petroleum gas under low steam to carbon molar ratios. *Int. J. Hydrogen Energy* **2010**, *35*, 12191–12200. [CrossRef]

28. Luisetto, I.; Tuti, S.; Battocchio, C.; Lo Mastro, S.; Sodo, A. Ni/CeO$_2$–Al$_2$O$_3$ catalysts for the dry reforming of methane: The effect of CeAlO$_3$ content and nickel crystallite size on catalytic activity and coke resistance. *Appl. Catal. A* **2015**, *500*, 12–22. [CrossRef]

29. Penkova, A.; Bobadilla, L.; Ivanova, S.; Domínguez, M.I.; Romero-Sarria, F.; Roger, A.C.; Centeno, M.A.; Odriozola, J.A. Hydrogen production by methanol steam reforming on NiSn/MgO–Al$_2$O$_3$ catalysts: The role of mgo addition. *Appl. Catal. A* **2011**, *392*, 184–191. [CrossRef]

30. Izquierdo, U.; Barrio, V.L.; Requies, J.; Cambra, J.F.; Guemez, M.B.; Arias, P.L. Tri-reforming: A new biogas process for synthesis gas and hydrogen production. *Int. J. Hydrogen Energy* **2013**, *38*, 7623–7631. [CrossRef]

31. Chattanathan, S.A.; Adhikari, S.; McVey, M.; Fasina, O. Hydrogen production from biogas reforming and the effect of H_2S on CH_4 conversion. *Int. J. Hydrogen Energy* **2014**, *39*, 19905–19911. [CrossRef]

Article

Catalytic Steam Reforming of Toluene: Understanding the Influence of the Main Reaction Parameters over a Reference Catalyst

Hua Lun Zhu, Laura Pastor-Pérez and Marcos Millan *

Department of Chemical Engineering, Imperial College London, London SW7 2AZ, UK; h.zhu16@imperial.ac.uk (H.L.Z.); l.pastor-perez@imperial.ac.uk (L.P.-P.)
* Correspondence: marcos.millan@imperial.ac.uk

Received: 31 December 2019; Accepted: 28 January 2020; Published: 13 February 2020

Abstract: Identifying the suitable reaction conditions is key to achieve high performance and economic efficiency in any catalytic process. In this study, the catalytic performance of a Ni/Al_2O_3 catalyst, a benchmark system—was investigated in steam reforming of toluene as a biomass gasification tar model compound to explore the effect of reforming temperature, steam to carbon (S/C) ratio and residence time on toluene conversion and gas products. An S/C molar ratio range from one to three and temperature range from 700 to 900 °C was selected according to thermodynamic equilibrium calculations, and gas hourly space velocity (GHSV) was varied from 30,600 to 122,400 h^{-1} based on previous work. The results suggest that 800 °C, GHSV 61,200 h^{-1} and S/C ratio 3 provide favourable operating conditions for steam reforming of toluene in order to get high toluene conversion and hydrogen productivity, achieving a toluene to gas conversion of 94% and H_2 production of 13 mol/mol toluene.

Keywords: toluene; steam reforming; GHSV; S/C ratio; coke formation

---

1. Introduction

Greenhouse gas (GHG) emissions from fossil fuel combustion for power generation represent a major contribution to climate change. For this very reason, a switch from conventional to renewable power resources, i.e., solar, wind, hydroelectric energy and biomass is necessary [1].

Biomass can consistently provide energy and fuels and has an advantage over other renewable energies sources as it is more homogeneously distributed over the earth and is an abundant resource [2]. International Energy Outlook 2017 reported that biomass could provide over 14% of the world's primary energy consumption, which is the highest among renewable energy resource, and it will contribute a quarter or third of the global primary energy supply by 2050 [3].

For all the above and as a consequence of unstable oil prices and the alarming climate change, biomass gasification has increasingly received interest [4]. Indeed, this is a versatile and interesting way to re-use different wastes (e.g., agricultural and urban wastes, energy crops, food and industrial processing residues) to produce bio-syngas, which can be used for electrical power generation (fuel cells, gas turbine or engine), or as feedstock for the synthesis of liquid fuels and chemicals such as methanol [5]. Furthermore, the necessary technology for this process can be adapted from old coal gasification units [6]. However, one of the most critical technical challenges in biomass gasification is the formation of tars. Tar condensation can cause serious risks to downstream equipment. Therefore, tars should be removed from the effluent stream of biomass gasification [7].

Existing techniques for tar removal after a gasifier include separation either by physical (mechanical) methods, using ceramic candle filters or wet scrubbers, or thermochemical conversion methods using high temperature thermal or catalytic cracking to convert tar into syngas [8]. Physical separation methods would cause secondary pollution since they only remove tar from gas products, resulting in a waste stream that needs treatment. Conversely, thermal cracking has received increasing attention because tar can be converted into useful gas products and increases the overall efficiency of the gasification process [9]. Thermal cracking without catalysts operates at high temperature (>1000 °C) to decompose the tars in smaller non-condensable molecules. The high energy consumption makes this process less interesting. By contrast, catalytic cracking of tars can be carried out at lower temperatures converting tars into useful gases in a more efficient manner and is being widely studied as a principal method for tar removal [10]. As a steam reforming reaction, the proposed reaction could remove tar by a catalytic process and produce fuel H_2 and CO at relatively low temperatures. At the same time, tar steam reforming poses some challenges that must be addressed related to the reaction conditions. Reforming catalysts can lose activity over time due to carbon deposition and sintering over the active phases [5]. These problems can be minimised with optimal operating conditions in the presence of the right catalyst. Real tar composition is highly complex and most studies use model tar compounds such as toluene, benzene or naphthalene to ascertain the catalytic mechanism [11–13]. However, previous studies have shown that these compounds represent a worst-case scenario in the tendency of the system to form carbon deposits [14,15].

Noble metals and Ni are most widely active phases used in reforming catalysts. Noble metal catalysts including Pt, Rh and Ru are known for their exceptionally good activity and stability in tar steam reforming. However, these catalysts have had limited use due to their high costs [16].

Nowadays, the aim is to develop an economically viable material, ideally not containing noble metals, which produces the same high levels of conversion and reaction performance as the noble ones. Ni is an attractive choice as steam reforming catalytic metal thanks to its good performance in the conversion of different types of hydrocarbons [16], being, for instance, the most popular active phase in methane steam reforming [17,18]. In particular, Ni/Al_2O_3 catalysts are considered as the state-of-the-art materials for steam reforming processes. Furthermore, the high surface area of alumina and its mechanical properties result in an excellent choice as support for nickel nanoparticles [19].

Under these premises, this work showcases the application of a Ni/Al_2O_3 catalyst in the steam reforming of toluene (C_7H_8) as a tar model compound in a fixed bed reactor. Until now, the catalytic performance in the steam reforming of toluene has been mainly evaluated as a function of catalyst design variables, such as the nature of the support [20–22] and metal [23,24], but little attention has been paid to the reaction conditions, especially for this benchmark catalytic formulation [25]. Identification and optimisation of the reaction parameters in the presence of a commercial-like catalyst (Ni/Al_2O_3) are vital to achieving the best catalytic performance. However, few studies involving parameter screening have been carried out [8].

Herein we analyse the influence of reforming temperature, steam to carbon molar ratio (S/C) and gas hourly space velocity (GHSV) on the toluene reforming performance. These are considered the key parameters to fine-tune the reaction and maximise the overall performance.

Parallel reactions in this system can be numerous. The main reactions that can occur during toluene steam reforming are represented as follows:

Toluene steam reforming

$$C_7H_8 + 14\,H_2O \rightarrow 7CO_2 + 18\,H_2 \tag{1}$$

$$C_7H_8 + 7\,H_2O \rightarrow 7CO + 11\,H_2 \tag{2}$$

Water–gas shift

$$CO + H_2O \leftrightarrow CO_2 + H_2 \tag{3}$$

Hydrodealkylation

$$C_7H_8 + H_2 \rightarrow C_6H_6 + CH_4 \tag{4}$$

Methane steam reforming

$$CH_4 + H_2O \leftrightarrow CO + 3H_2 \tag{5}$$

Boudouard reaction

$$2CO \leftrightarrow CO_2 + C \tag{6}$$

Steam reforming of toluene is irreversible and Reactions (1) and (2) are dependent on the S/C ratio used. CH_4 is produced from hydroalkylation (Reaction (4)) and pyrolysis of toluene. Methane reforming followed by a water–gas shift reaction converts the produced CO with steam to H_2 and CO_2. A Boudouard reaction is an exothermic reaction that produces C from CO. All of these reactions are heavily conditioned by the temperature, space velocity and reactants ratio [26–28]. Hence, a careful assessment of the impact of these parameters will allow us to identify the optimum conditions towards the generation of added value products.

For all the above, this work fills an essential gap in tar reforming literature via the systematic study of key reaction parameters using a state-of-the-art catalyst to reveal the optimum process conditions.

2. Experimental

2.1. Catalyst Preparation

The nickel-based catalyst was prepared following a wet impregnation method. The necessary amount of $Ni(NO_3)_2 \cdot 6H_2O$ (≥97.0%, Sigma-Aldrich) to obtain 20 wt.% NiO was dissolved in an excess of acetone (≥99.8%, Sigma Aldrich). Then, the support γ- Al_2O_3 (≥98.0% purity, Sasol) was added into the solution and, after stirring for 2 h, the solvent was removed under vacuum at 60 °C by using a rotating evaporator. The remaining mixture was dried overnight at 110 °C. Finally, the solid was calcined at 600 °C with a ramping rate of 2 °C·min^{-1} for 4 h. It has been reported that Ni/Al_2O_3 catalysts are stable under reaction conditions despite the calcination temperature being lower than those of the experiments [29,30]. Lower calcination temperatures have been shown to lead to better catalytic performance in steam reforming [31]. The obtained sample was labelled Ni/Al_2O_3. The Ni content assuming complete reduction from NiO to Ni is 16.4 wt.%.

2.2. Characterisation

Thermogravimetric analysis (TGA) was carried out to investigate the coke deposition on the catalyst in a Pyris 1 thermogravimetric analyser from PerkinElmer (Waltham, MA, USA). The samples were ramped from room temperature to 900 °C at a rate of 10 °C·min^{-1} in air.

N_2-adsorption-desorption analysis was conducted in a TriStar 3000 V6.07 A analyser from Micromeritics (Norcross, GA, USA). Before the analysis, the catalyst was degassed at 150 °C for 4 h in a vacuum. The Brunauer–Emmett–Teller (BET) method was used to calculate the surface area of the catalyst.

2.3. Catalytic Toluene Steam Reforming Tests

Toluene steam reforming was carried out in a fixed bed reactor used in previous bio-oil reforming work [24]. Before the reaction, the reactor was purged with N_2 to remove air. The catalyst was reduced under 50 mL·min^{-1} of H_2 up to 700 °C for 1 h before each test.

Figure 1 shows the schematic diagram of the experimental reaction set-up, and the reaction zone is shown in Figure 2. The reactor was heated up by two copper electrodes; toluene and steam were injected by two syringe pumps from the top of the reactor and preheated at 200 °C to the vapour phase in a preheating chamber. Toluene was carried by N_2 with a fixed concentration of 100 g Nm^{-3}. Then the reactant stream entered an incoloy alloy 625 tube (12 mm i.d., 2 mm thick, 253 mm long), equipped with an inner quartz tube (9 mm i.d., 1 mm thick, 300 mm long) to prevent any contact between the reactant gas stream and the incoloy internal surface. 500 mg of Ni/Al_2O_3 catalyst with a particle size in the range of 250–500 µm was placed right in the middle of the quartz tube. A K-type thermocouple was used to determine the catalytic bed temperature.

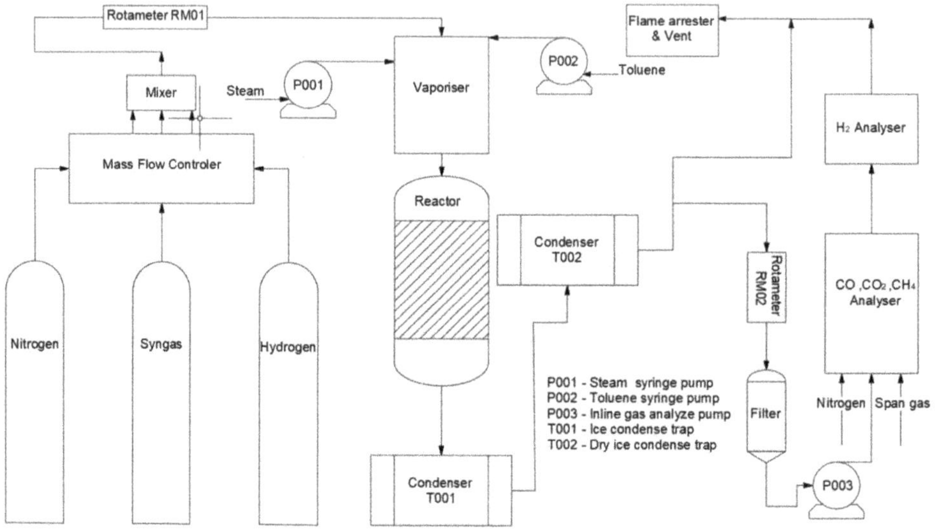

Figure 1. Schematic diagram of the catalytic toluene steam reforming system.

The product gases after reaction pass through two condensers in series to collect any liquid product as well as unreacted toluene and water. Ice and dry ice were used as coolant in the two condensers, respectively. The products identified in the gas phase were H_2, CH_4, CO_2 and CO. Two on-line gas analysers were used to determine the product gas compositions: an MGA3000 Multi-Gas infrared analyser (ADC Gas Analysis, Herts, UK) for CO_2, CH_4 and CO, followed by a K1550 thermal conductivity H_2 analyser (Hitech Instruments, Luton, UK).

The performance of catalysts was evaluated by the conversion into gaseous products (based on a carbon balance between the inlet and the outlet stream of the reactor), selectivity to main products (where "i" is CO_2, CO and CH_4 in moles) and hydrogen yield, which were defined as follows:

$$\% \text{ Carbon Conversion} = \frac{\text{C in the gas product}}{\text{C fed into reactor}} * 100 \tag{7}$$

$$\% \text{ "i" selectivity} = \frac{\text{"i" produced in moles}}{\text{C atoms in the gas products}} * 100 \tag{8}$$

$$\text{Hydrogen yield} = \frac{\text{total Hydrogen production in moles}}{\text{toluene fed into reactor in moles}} \tag{9}$$

The experimental error in toluene conversion, gas selectivity and gas yield is ±2%. Toluene conversion and H_2 production could be influenced by experiment conditions and parameters. Reforming temperature, S/C ratio and residence time are reported to be the key factors that would affect the total conversion and H_2 yield. In this paper S/C ratios of 1, 2 and 3; temperatures of 700, 800 and 900 °C, and GHSV of 30,600, 61,200, 91,800 and 122,400 h^{-1} were investigated.

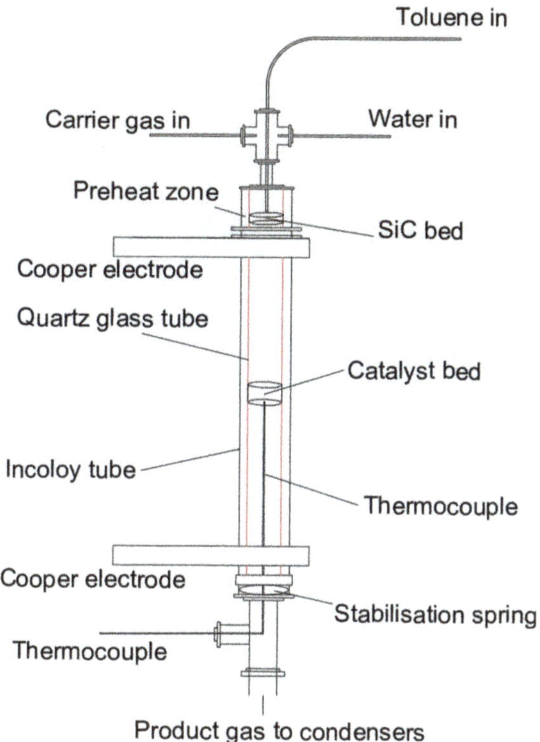

Figure 2. Reactor diagram of catalytic toluene steam reforming.

2.4. Thermodynamic Simulation

The ASPEN software package (AspenTech, Bedford, MA, USA) was used to determine the thermodynamic equilibrium of the toluene reforming reactions over the different reaction conditions. An ideal property method, an RGIBBS reactor (based on Gibbs free energy minimisation) was selected to investigate the thermodynamic equilibrium. Material flows into the reactor are identical to those from the corresponding experiment. The influence and effects of experimental parameters, including reforming the temperature and S/C ratio on the toluene conversion, the yield of main light gases and the carbon deposition was investigated.

3. Results and Discussion

3.1. Textural Properties of the Synthetised Catalyst

The N_2 adsorption-desorption isotherm of the calcined Ni/Al_2O_3 catalyst is shown in Figure 3, which shows a type IV isotherm with a characteristic hysteresis loop for mesoporous materials. The adsorption average pore width was 9.1 nm, the total pore volume was 0.35 $cm^3 \cdot g^{-1}$ and BET Surface Area was 153 $m^2 \cdot g^{-1}$. The BET surface area, pore volume and pore width of the original Al_2O_3 was 230 $m^2 \cdot g^{-1}$, 0.5 $cm^3 \cdot g^{-1}$ and 10 nm respectively. Textural properties are actually governed by the primary support gamma-alumina, which provides mechanical and thermal stability as well as high surface area.

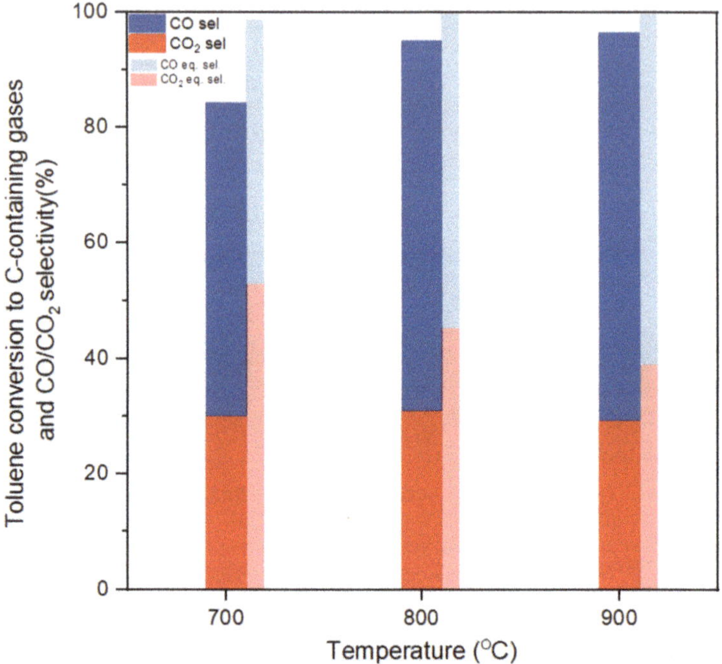

Figure 3. Toluene conversion to C-containing gases and CO/CO$_2$ selectivity at different temperatures (S/C:3, gas hourly space velocity (GHSV):61,200 h^{-1}).

3.2. Influence of Temperature on Toluene Steam Reforming

The steam reforming temperature was concluded to have a significant influence on toluene conversion since higher temperatures could increase syngas production and conversion from toluene to gas products [32]. When reforming is envisaged as a downstream tar upgrading unit after gasification, the most interesting temperature interval for atmospheric reforming is between 600 and 900 °C, since the gasification effluent temperature will normally be lower than 900 °C [33]. Research also suggested that high temperature might reduce H$_2$ yield as the reverse water–gas shift reaction is favoured due to its endothermic nature [34–36]. The experiments were conducted at different temperatures to investigate the most suitable conditions for toluene conversion and H$_2$ production. Catalytic performance tests were performed at 700, 800 and 900 °C, with a fixed S/C ratio of 3 and a GHSV of 61,200 h^{-1} (corresponding to a N$_2$ flow rate of 300 mL·min^{-1}) for 5 h.

Figure 3 shows the toluene conversion and CO$_2$/CO selectivity (based on conversion to C-containing gases) at the three different temperatures in a steady state within the five hours of reaction and compares with equilibrium selectivity. CO and CO$_2$ were the main gases. The experimental selectivity of CH$_4$ at all temperatures was under 0.1%. Conversion and selectivity of the gases approached that of thermodynamic equilibrium as temperature increased. Thermodynamic equilibrium predicts total toluene conversion at 700 °C or higher temperatures. At 700 °C, the thermodynamic equilibrium indicated a CH$_4$ selectivity of 1.1%, and for higher temperatures only CO and CO$_2$ were predicted as C-containing gas products. Experimental toluene conversions ranged from 84% to 96% and approached equilibrium with the temperature increase from 700 to 900 °C. Coke formation only accounted for less than 1% of toluene conversion, the rest being to gas phase products. Unreacted toluene was condensed in the cooling trap, which enabled closing the mass balance within 98%. The high conversions of toluene achieved for the three temperatures highlight once again the good performance of this commercial-like catalyst (Ni/Al_2O_3). Both experimental and equilibrium selectivity showed that the CO/CO_2 ratio increases as temperature increases. This is likely dominated by a reverse water–gas shift reaction in the high-temperature area (Reaction (3), as the increasing temperature would favour the endothermic direction, and nickel contents would also promote the reaction at high temperatures [11].

Table 1 shows the gaseous product yields including CO, CO_2, H_2 and CH_4. The yields of CO and CO_2 have similar trends to selectivity. However, for H_2 yield the maximum was achieved at 800 °C with a H_2 production of 13.0 mol/mol toluene, while the highest H_2 production in equilibrium conditions was predicted at 700 °C. This can be due to, in the experimental test at 700 °C, the lowest toluene conversion into gases was obtained. Toluene conversion to gas stayed over 94% at 800 °C or above, while a higher temperature would lead to a slight decrease in the content of H_2 in gaseous products due to the presence of the reverse water–gas shift reaction as discussed above. It is interesting to remark that only at 700 °C the undesirable CH_4 side product was obtained and it was only in a small amount. The absence of CH_4 at higher temperatures can be due to the methane reforming to CO and other parallel processes consuming methane as suggested by the equilibrium results.

Table 1. Product yields for the gaseous products at the three different temperatures (S/C:3, GHSV: 61,200 h^{-1}).

Temperature (°C)	CO_2 (mol/mol Toluene)	CO (mol/mol Toluene)	H_2 (mol/mol Toluene)	CH_4 (mol/mol Toluene)
700	2.1	3.8	10.3	0
(equilibrium)	(3.7)	(3.2)	(14.5)	(0.1)
800	2.2	4.5	13.0	0
(equilibrium)	(3.2)	(3.8)	(14.3)	(0)
900	2.1	4.7	11.8	0
(equilibrium)	(2.8)	(4.2)	(13.8)	(0)

Thermogravimetric analysis was conducted on the catalysts after a five-hour reaction to estimate the coke formation during the three different temperatures reaction tests. It has been reported that coke deposition on Ni/Al_2O_3 catalysts is the main cause for deactivation and high nickel contents could also favour coke formation. Although coke deposition is thermodynamically unfavourable at high temperatures (>600 °C), methane decomposition in the high temperature range could lead to the production of solid carbon [37]. Table 2 shows the carbon conversion from toluene to coke and the fraction of coke on the catalyst as a function of temperature. As the temperature increased, the conversion to carbon deposits was slightly higher. It is likely that carbon at the higher temperatures was formed from the reforming of CH_4 observed at lower temperatures, as CH_4 is known to favour coke formation on Ni-based catalysts [27]. Notwithstanding coke formation at these temperatures, the amount of coke is very low, and the catalyst remains stable during the whole experiment.

Table 2. Toluene conversion to coke and fraction of coke deposited on the catalyst at different reforming temperatures (S/C:3, GHSV:61,200 h^{-1}, 5-h test).

Reforming Temperature	700 °C	800 °C	900 °C
Coke/C in toluene	0.43%	0.61%	0.77%
Coke/Catalyst (g_C/g_{cat})	0.118	0.167	0.211

Based on the results presented above, 800 °C was considered the most suitable reforming temperature due to the highest H_2 production, an excellent overall conversion (94%) and acceptable coke deposition levels. On the contrary, despite the 900 °C experiment showing a slightly higher toluene conversion, lower H_2 production, greater coke deposition and lower energy efficiency made this temperature not preferable with the 800 °C test.

Gas analysis as a function of time on stream at the chosen temperature is shown in Figure 4 for a five-hour test. There was no CH_4 detected throughout the test and CO, CO_2 and H_2 concentration stayed stable *ca* 11%, 23% and 66% (figures corrected from N_2 dilution), respectively, during the whole experiment. Furthermore, no obvious change in the CO/CO_2 ratio, or drop in H_2 yields and catalyst deactivation were observed in this test.

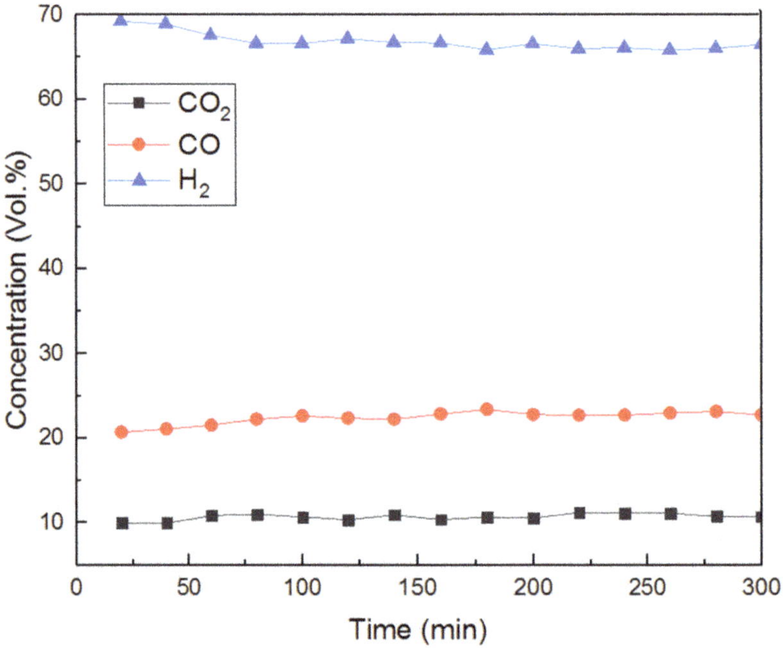

Figure 4. Gas product concentration at 800 °C, S/C:3, GHSV:61,200 h^{-1} in a 5-h test.

3.3. Influence of GHSV in Toluene Steam Reforming

In view of the results obtained in the temperature screening, a temperature of 800 °C was chosen to study the effect of GHSV to elucidate the suitable residence time for a complete conversion of toluene. High GHSV, or lower residence time, might inhibit toluene conversion and H_2 production. Literature suggests that suitable GHSVs for steam reforming were mostly between 10,000 and 70,000 h^{-1} [38,39] over different catalysts. For our benchmark catalyst, tests were carried out with 500 mg of catalyst, and a fixed temperature (800 °C) and S/C ratio of three during 5 h of reaction. The feeding rates of toluene, steam and N_2 were changed according to the desired GHSV. The GHSV studied were 30,600, 61,200, 91,800 and 122,400 h^{-1}. Some of the space velocities selected in this study are indeed over the standard ranges mentioned above. In fact, high space velocities normally involve lower reactor volumes, thus decreasing the capital cost of the reformer.

Figure 5 shows the toluene conversion and CO_2/CO selectivity (based on conversion to C-containing gases) at the four different GHSVs in a steady state within the five hours of reaction and compares them with equilibrium selectivity. Toluene conversion decreased from 96% to 86% when GHSVs increased from 30,600 to 122,400 h^{-1}. When GHSV was 91,800 h^{-1} or lower, toluene conversion to gas remained higher than 90%, without notorious differences for the two lowest GHSV studied (30,600 h^{-1} or 61,200 h^{-1}), which showed a stabilised conversion at around 95%. The selectivity of CO and CO_2 approached the equilibrium results slowly when GHSV decreased. It could be seen that toluene conversion and CO_2 selectivity declined with the increasing of GHSV, as a result of shorter residence time and reaction period. It is interesting to note that the selectivity is less affected by the residence time than the conversion. Negligible or no methane was detected in the product gas, which was in line with literature that suggested that at temperatures higher than 750 °C very small amounts of CH_4 are produced [11].

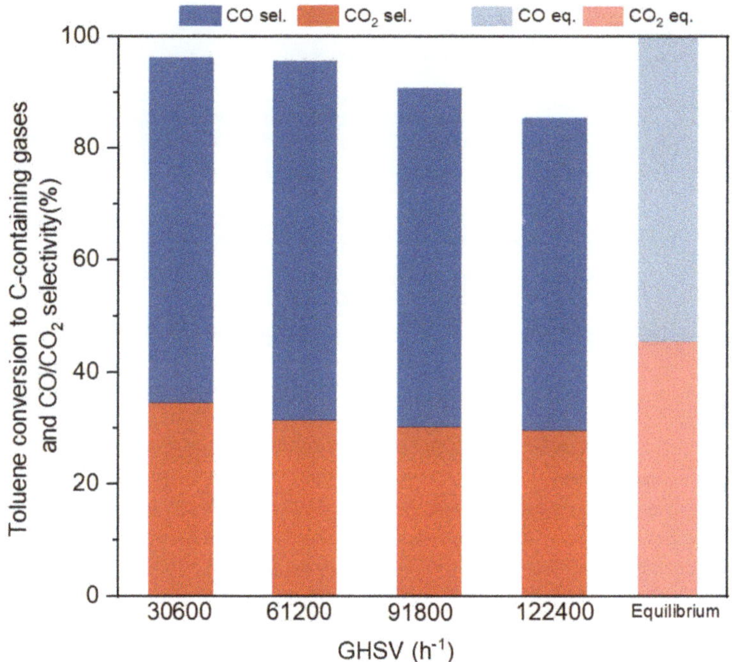

Figure 5. Toluene conversion to C-containing gases and CO/CO_2 selectivity at different GHSVs (S/C:3, 800 °C).

The gas product CO_2, CO and H_2 yields are shown in Table 3. As expected, GHSV 30,600 h^{-1} led to the highest H_2 production (13.2 mol/mol toluene). As a trend, CO, CO_2 and H_2 yields dropped when GHSV increased. This decrease in yields for all gas products at high GHSV is linked with the drop in conversion as the residence time became shorter. An exception occurred at 61,200 h^{-1} GHSV as the reverse water–gas shift reaction could cause a small increase in CO. Previous studies suggest that some tar model compounds (naphthalene) had no apparent trends for the hydrogen yields or selectivity, as the product was affected by the equilibrium of the water–gas shift reaction and other side reactions [40]. In this work, hydrogen yield and total conversion showed a slightly increasing trend in conversion and yields with the decrease of residence time, but this trend was more remarkable in the conversion of toluene than in selectivities.

Table 3. Product yields for the gaseous products at different GHSV (S/C:3, 800 °C).

GHSV (h^{-1})	CO_2 (mol/mol Toluene)	CO (mol/mol Toluene)	H_2 (mol/mol Toluene)
30,600	2.4	4.3	13.2
61,200	2.2	4.5	13.0
91,800	2.1	4.2	12.8
122,400	2.1	3.9	10.8
(Equilibrium)	(3.2)	(3.8)	(14.3)

Table 4 shows the carbon conversion from toluene to coke with different GHSVs. This data led us to a better understanding of the coke deposition. The reactant toluene and steam feeding rate increased three times when GHSV increased from 30,600 to 122,400 h^{-1}; however, the coke conversion reduced from 0.38% to 0.22%. This means that the higher GHSV used resulted in a higher coke amount deposited but lower conversion rate. This result is the balance between the higher feed of toluene used and the decrease in conversion due to the high GHSV used.

Table 4. Toluene conversion to coke deposited on the catalyst with different GHSV (S/C:3, 800 °C, five-hour test).

GHSV(h^{-1})	30,600	61,200	91,800	122,400
Coke/C in toluene	0.38%	0.31%	0.23%	0.22%
Coke/Catalyst (g_C/g_{cat})	0.105	0.167	0.188	0.242

Although the lower space velocity leads to higher conversion, greater selectivity and the lower net amount of carbon deposits, for practical applications (i.e. manufacturing cost savings) higher space velocities are desired. In this sense, 61,200 h^{-1} yields very similar conversion and selectivity levels, low net coking and lower conversion to coke. Hence, this space velocity was selected to optimise the next reaction parameter.

3.4. Influence of S/C Ratio in Toluene Steam Reforming

Steam, as a principal reactant in catalytic steam reforming, is widely recognised to have a strong influence on H_2 production. Steam is involved in most of the relevant reactions in the toluene reforming and therefore the steam to carbon ratio was chosen as a key variable to study. A high steam partial pressure improves gasification reactions and moves the water–gas shift equilibrium towards hydrogen production, while the partial pressure of toluene in the gas stream is lower due to dilution as the S/C ratio rises [37]. Suitable S/C ratios to investigate the catalyst performance were mostly between one and four in the literature [16,41]. Although steam is the main reactant in a reforming process, a large excess of water in these experiments could condense into ice in the cooling trap and block the system. It is also reported that the saturation of the catalyst surface by steam at a high S/C ratio would not favour the conversion of toluene or H_2 production [5].

In this study, the catalytic performance tests were performed at three different S/C ratios of 1, 2, 3, at 800 °C, 500 mg of samples and a GHSV of 61,200 h^{-1} for five hours.

Figure 6 shows the influence of the S/C ratio on the selectivity of carbon containing products. Both experimental and equilibrium results showed that an increment in S/C ratio results in a large improvement of CO_2 selectivity and toluene conversion. In the experimental tests, toluene to gas conversion increased from 53% to 94% when the S/C ratio increased from one to three. Equilibrium results indicated that a higher S/C ratio always increases the H_2 production, as the excess steam would promote the water–gas shift reaction. A higher S/C ratio also increased toluene conversion to gas. Some researchers suggested that the most suitable S/C ratio was mostly between 2.5 and 3.5, because more excess steam would not increase toluene conversion or H_2 production, causing a drop in toluene partial pressure [26].

Table 5 compares the product yields of CO, CO_2 and H_2 with different S/C ratios. Equilibrium simulation and experiments both showed large increases in CO_2 and H_2 production with the S/C ratio. Experimental H_2 yield increased from 5.1 to 13.0 mol/mol toluene when the S/C ratio raised from one to three. The opposite equilibrium trend is expected for CO, but this did not happen in the experiments. Instead, the CO yield was observed to increase with the amount of steam as the trend was dominated by the higher conversion achieved at higher S/C ratios. Despite the experimental CO yield showing a slight increase, in general the CO/CO_2 ratio decreased with the increase of the S/C ratio in line with the equilibrium results.

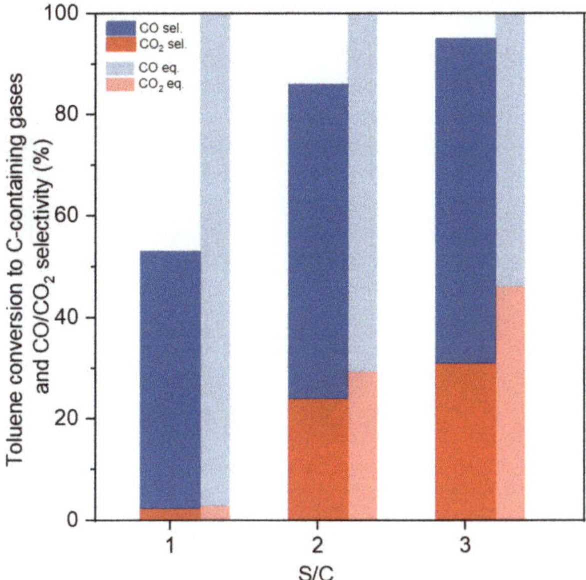

Figure 6. Toluene conversion to C-containing gases and CO/CO_2 selectivity at different S/C ratios (GHSV:61,200 h^{-1}, 800 °C).

Table 5. Product yields for the gaseous products under different S/C ratios (GHSV:61,200 h^{-1}, 800 °C).

S/C ratio	CO_2 (mol/mol Toluene)	CO (mol/mol Toluene)	H_2 (mol/mol Toluene)
1	0.2	3.5	5.1
(Equilibrium)	(0.2)	(6.8)	(9.8)
2	1.7	4.3	10.5
(Equilibrium)	(2.1)	(4.9)	(13.0)
3	2.2	4.5	13.0
(Equilibrium)	(3.2)	(3.8)	(14.3)

The coke weight on the spent catalyst with S/C ratio 1, 2, 3 is 0.289, 0.222, 0.167 g$_C$/g$_{cat}$, respectively. As expected, the higher S/C ratio promoted carbon gasification avoiding coke formation [42]. Table 6 shows the carbon conversion from toluene to coke for the three different S/C ratio studied. As discussed above, it can be seen that the excess of steam inhibited, but did not suppress, the formation of coke. For all the above, the better choice is to use a S/C ratios of three since it permits the highest H_2 production and toluene conversion and the lowest coke amount.

Table 6. Toluene conversion to coke deposited on the catalyst at different S/C ratios.

S/C ratio	1	2	3
Coke/C in toluene	1.07%	0.82%	0.61%
Coke/Catalyst (g$_C$/g$_{cat}$)	0.289	0.222	0.167

4. Conclusions

To remove tar produced from biomass gasification, catalytic steam reforming was conducted for toluene as a model tar compound. Simulations of a thermodynamic equilibrium based on Gibbs free energy minimisation and experiments in a fixed bed reactor using a Ni/Al$_2$O$_3$ catalyst were carried out. The effect of reforming temperature, S/C ratio and GHSV on toluene conversion and product distribution was studied.

Increasing the temperature from 700 to 900 °C increased total conversion, with a potential risk of higher coke deposition. A temperature of 800 °C observed the highest H_2 production, high toluene conversion (>94%) and relatively lower coke deposition.

The Ni/Al_2O_3 catalyst only requires a very short residence time (GHSV < 91,800 h^{-1}) for toluene reforming with this catalyst and effectively removes low density toluene in a mixed gas stream. H_2 yield and toluene conversion increased slightly and approached simulated thermodynamic equilibrium results when GHSV decreased. Coke deposition increased at a lower rate as GHSV increased.

A high S/C ratio would greatly increase total conversion, hydrogen production and reduce the coke formation on the catalyst. The presence of excess steam could shift the equilibrium of the water–gas shift reaction to produce more H_2.

A temperature of 800 °C, GHSV of 61,200 h^{-1} and S/C ratio of three provided the most suitable reaction conditions for toluene conversion and H_2 production in steam reforming of toluene, obtained a steady state of a toluene to gas conversion over 94%, a H_2 production of 141.6 mol/mol toluene in a five-hour test, with no obvious deactivation observed in five hours. Based on these results, this condition would be suitable for tar model compound removal.

Author Contributions: Conceptualization, M.M. and H.L.Z.; experiments, H.L.Z.; data curation and analysis, H.L.Z. and L.P.-P.; writing—original draft preparation, H.L.Z.; writing—review and editing, M.M. and L.P.-P.; supervision, M.M. and L.P.-P.; All authors have read and agreed to the published version of the manuscript.

Funding: This research received no external funding.

Conflicts of Interest: The authors declare no conflict of interest.

References

1. Farzad, S.; Mandegari, M.A.; Görgens, J.F. A critical review on biomass gasification, co-gasification, and their environmental assessments. *Biofuel Res. J.* **2016**, *3*, 483–495. [CrossRef]
2. Göransson, K.; Söderlind, U.; He, J.; Zhang, W. Review of syngas production via biomass DFBGs. *Renew. Sustain. Energy Rev.* **2011**, *15*, 482–492. [CrossRef]
3. International Energy Agency. *International Energy Outlook 2017*; International Energy Agency: Paris, France, 2017.
4. Balat, H.; Kırtay, E. Hydrogen from biomass–present scenario and future prospects. *Int. J. Hydrogen Energy* **2010**, *35*, 7416–7426. [CrossRef]
5. Świerczyński, D.; Libs, S.; Courson, C.; Kiennemann, A. Steam reforming of tar from a biomass gasification process over Ni/olivine catalyst using toluene as a model compound. *Appl. Catal. B-Environ.* **2007**, *74*, 211–222. [CrossRef]
6. Claude, V.; Mahy, J.G.; Douven, S.; Pirard, S.L.; Courson, C.; Lambert, S.D. Ni-and Fe-doped γ-Al_2O_3 or olivine as primary catalyst for toluene reforming. *Mater. Today Chem.* **2019**, *14*, 100197. [CrossRef]
7. Devi, L.; Ptasinski, K.J.; Janssen, F.J. A review of the primary measures for tar elimination in biomass gasification processes. *Biomass Bioenergy* **2003**, *24*, 125–140. [CrossRef]
8. Anis, S.; Zainal, Z.A. Tar reduction in biomass producer gas via mechanical, catalytic and thermal methods: A review. *Renew. Sustain. Energy Rev.* **2011**, *15*, 2355–2377. [CrossRef]
9. Noichi, H.; Uddin, A.; Sasaoka, E. Steam reforming of naphthalene as model biomass tar over iron–aluminum and iron–zirconium oxide catalyst catalysts. *Fuel Process. Technol.* **2010**, *91*, 1609–1616. [CrossRef]
10. Guan, G.; Kaewpanha, M.; Hao, X.; Abudula, A. Catalytic steam reforming of biomass tar: Prospects and challenges. *Renew. Sustain. Energy Rev.* **2016**, *58*, 450–461. [CrossRef]
11. Zhao, B.; Zhang, X.; Chen, L.; Qu, R.; Meng, G.; Yi, X.; Sun, L. Steam reforming of toluene as model compound of biomass pyrolysis tar for hydrogen. *Biomass Bioenergy* **2010**, *34*, 140–144. [CrossRef]
12. Virginie, M.; Courson, C.; Kiennemann, A. Toluene steam reforming as tar model molecule produced during biomass gasification with an iron/olivine catalyst. *Comptes Rendus Chimie* **2010**, *13*, 1319–1325. [CrossRef]

13. Park, H.J.; Park, S.H.; Sohn, J.M.; Park, J.; Jeon, J.K.; Kim, S.S.; Park, Y.K. Steam reforming of biomass gasification tar using benzene as a model compound over various Ni supported metal oxide catalysts. *Bioresour. Technol.* **2010**, *101*, S101–S103. [CrossRef] [PubMed]

14. Lorente, E.; Millan, M.; Brandon, N.P. Use of gasification syngas in SOFC: Impact of real tar on anode materials. *Int. J. Hydrogen Energy* **2012**, *37*, 7271–7278. [CrossRef]

15. Lorente, E.; Berrueco, C.; Millan, M.; Brandon, N.P. Effect of tar fractions from coal gasification on nickel-yttria stabilized zirconia and nickel-gadolinium doped ceria solid oxide fuel cell anode materials. *J. Power Sources* **2013**, *242*, 824–831. [CrossRef]

16. Coll, R.; Salvado, J.; Farriol, X.; Montane, D. Steam reforming model compounds of biomass gasification tars: Conversion at different operating conditions and tendency towards coke formation. *Fuel Process. Technol.* **2001**, *74*, 19–31. [CrossRef]

17. Matsumura, Y.; Nakamori, T. Steam reforming of methane over nickel catalysts at low reaction temperature. *Appl. Catal. A-Gen.* **2004**, *258*, 107–114. [CrossRef]

18. Lertwittayanon, K.; Youravong, W.; Lau, W.J. Enhanced catalytic performance of Ni/α-Al$_2$O$_3$ catalyst modified with CaZrO$_3$ nanoparticles in steam-methane reforming. *Int. J. Hydrogen Energy* **2017**, *42*, 28254–28265. [CrossRef]

19. Ahmed, T.; Xiu, S.; Wang, L.; Shahbazi, A. Investigation of Ni/Fe/Mg zeolite-supported catalysts in steam reforming of tar using simulated-toluene as model compound. *Fuel* **2018**, *211*, 566–571. [CrossRef]

20. Mermelstein, J.; Millan, M.; Brandon, N.P. The impact of carbon formation on Ni–YSZ anodes from biomass gasification model tars operating in dry conditions. *Chem. Eng. Sci.* **2009**, *64*, 492–500. [CrossRef]

21. Xu, H.; Liu, Y.; Sun, G.; Kang, S.; Wang, Y.; Zheng, Z.; Li, X. Synthesis of graphitic mesoporous carbon supported Ce-doped nickel catalyst for steam reforming of toluene. *Mater. Lett.* **2019**, *244*, 123–125. [CrossRef]

22. Li, Z.; Li, M.; Ashok, J.; Kawi, S. NiCo@ NiCo phyllosilicate@ CeO$_2$ hollow core shell catalysts for steam reforming of toluene as biomass tar model compound. *Energy Convers. Manag.* **2019**, *180*, 822–830. [CrossRef]

23. Ashok, J.; Dewangan, N.; Das, S.; Hongmanorom, P.; Wai, M.H.; Tomishige, K.; Kawi, S. Recent progress in the development of catalysts for steam reforming of biomass tar model reaction. *Fuel Process. Technol.* **2020**, *199*, 106252. [CrossRef]

24. Bizkarra, K.; Bermudez, J.M.; Arcelus-Arrillaga, P.; Barrio, V.L.; Cambra, J.F.; Millan, M. Nickel based monometallic and bimetallic catalysts for synthetic and real bio-oil steam reforming. *Int. J. Hydrogen Energy* **2018**, *43*, 11706–11718. [CrossRef]

25. Meng, J.; Zhao, Z.; Wang, X.; Wu, X.; Zheng, A.; Huang, Z.; Li, H. Effects of catalyst preparation parameters and reaction operating conditions on the activity and stability of thermally fused Fe-olivine catalyst in the steam reforming of toluene. *Int. J. Hydrogen Energy* **2018**, *43*, 127–138. [CrossRef]

26. Oh, G.; Park, S.Y.; Seo, M.W.; Kim, Y.K.; Ra, H.W.; Lee, J.G.; Yoon, S.J. Ni/Ru–Mn/Al$_2$O$_3$ catalysts for steam reforming of toluene as model biomass tar. *Renew. Energy* **2016**, *86*, 841–847. [CrossRef]

27. Meidanshahi, V.; Bahmanpour, A.M.; Iranshahi, D.; Rahimpour, M.R. Theoretical investigation of aromatics production enhancement in thermal coupling of naphtha reforming and hydrodealkylation of toluene. *Chem. Eng. Process* **2011**, *50*, 893–903. [CrossRef]

28. Tang, Y.; Liu, J. Effect of anode and Boudouard reaction catalysts on the performance of direct carbon solid oxide fuel cells. *Int. J. Hydrogen Energy* **2010**, *35*, 11188–11193. [CrossRef]

29. Zhang, Y.; Williams, P.T. Carbon nanotubes and hydrogen production from the pyrolysis catalysis or catalytic-steam reforming of waste tyres. *J. Anal. Appl. Pyrol.* **2016**, *122*, 490–501. [CrossRef]

30. Yue, B.; Wang, X.; Ai, X.; Yang, J.; Li, L.; Lu, X.; Ding, W. Catalytic reforming of model tar compounds from hot coke oven gas with low steam/carbon ratio over Ni/MgO–Al2O3 catalysts. *Fuel Process. Technol.* **2010**, *91*, 1098–1104. [CrossRef]

31. Dieuzeide, M.L.; Iannibelli, V.; Jobbagy, M.; Amadeo, N. Steam reforming of glycerol over Ni/Mg/γ-Al2O3 catalysts. Effect of calcination temperatures. *Int. J. Hydrogen Energy* **2012**, *37*, 14926–14930. [CrossRef]

32. Bona, S.; Guillén, P.; Alcalde, J.G.; García, L.; Bilbao, R. Toluene steam reforming using coprecipitated Ni/Al catalysts modified with lanthanum or cobalt. *Chem. Eng. J.* **2008**, *137*, 587–597. [CrossRef]

33. Quitete, C.P.; Bittencourt, R.C.P.; Souza, M.M. Steam reforming of tar using toluene as a model compound with nickel catalysts supported on hexaaluminates. *Appl. Catal. A-Gen.* **2014**, *478*, 234–240. [CrossRef]

34. Yang, L.; Pastor-Pérez, L.; Gu, S.; Sepúlveda-Escribano, A.; Reina, T.R. Highly efficient Ni/CeO$_2$-Al$_2$O$_3$ catalysts for CO$_2$ upgrading via reverse water-gas shift: Effect of selected transition metal promoters. *Appl. Catal. B-Environ.* **2018**, *232*, 464–471. [CrossRef]

35. Zhu, H.L.; Zhang, Y.S.; Materazzi, M.; Aranda, G.; Brett, D.J.; Shearing, P.R.; Manos, G. Co-gasification of beech-wood and polyethylene in a fluidized-bed reactor. *Fuel Process. Technol.* **2019**, *190*, 29–37. [CrossRef]

36. Stroud, T.; Smith, T.J.; Le Saché, E.; Santos, J.L.; Centeno, M.A.; Arellano-Garcia, H.; Odriozola, J.A.; Reina, T.R. Chemical CO$_2$ recycling via dry and bi reforming of methane using Ni-Sn/Al$_2$O$_3$ and Ni-Sn/CeO$_2$-Al$_2$O$_3$ catalysts. *Appl. Catal. B-Environ.* **2018**, *224*, 125–135. [CrossRef]

37. Mukai, D.; Tochiya, S.; Murai, Y.; Imori, M.; Hashimoto, T.; Sugiura, Y.; Sekine, Y. Role of support lattice oxygen on steam reforming of toluene for hydrogen production over Ni/La$_{0.7}$Sr$_{0.3}$AlO$_3-\delta$ catalyst. *Appl. Catal. A-Gen.* **2013**, *453*, 60–70. [CrossRef]

38. Frusteri, F.; Freni, S.; Spadaro, L.; Chiodo, V.; Bonura, G.; Donato, S.; Cavallaro, S. H2 production for MC fuel cell by steam reforming of ethanol over MgO supported Pd, Rh, Ni and Co catalysts. *Catal. Commun.* **2004**, *5*, 611–615. [CrossRef]

39. He, L.; Hu, S.; Jiang, L.; Liao, G.; Chen, X.; Han, H.; Xiao, L.; Ren, Q.; Wang, Y.; Su, S.; et al. Carbon nanotubes formation and its influence on steam reforming of toluene over Ni/Al$_2$O$_3$ catalysts: Roles of catalyst supports. *Fuel Process. Technol.* **2018**, *176*, 7–14. [CrossRef]

40. Li, Q.; Wang, Q.; Kayamori, A.; Zhang, J. Experimental study and modeling of heavy tar steam reforming. *Fuel Process. Technol.* **2018**, *178*, 180–188. [CrossRef]

41. Li, C.; Hirabayashi, D.; Suzuki, K. Development of new nickel based catalyst for biomass tar steam reforming producing H2-rich syngas. *Fuel Process. Technol.* **2009**, *90*, 790–796. [CrossRef]

42. Le Saché, E.; Pastor-Pérez, L.; Garcilaso, V.; Watson, D.; Centeno, M.A.; Odriozola, J.A.; Reina, T.R. Flexible syngas production using a La$_2$Zr$_2$-xNixO$_7$-δ pyrochlore-double perovskite catalyst: Towards a direct route for gas phase CO$_2$ recycling. *Catal. Today* **2019**. [CrossRef]

Article

Application of Upgraded Drop-In Fuel Obtained from Biomass Pyrolysis in a Spark Ignition Engine

Alberto Veses [1],*, Juan Daniel Martínez [2], María Soledad Callén [1], Ramón Murillo [1] and Tomás García [1]

[1]   Instituto de Carboquímica, ICB-CSIC, Miguel Luesma Castán 4, 50018 Zaragoza, Spain; marisol@icb.csic.es (M.S.C.); ramon.murillo@csic.es (R.M.); tomas@icb.csic.es (T.G.)

[2]   Grupo de Investigaciones Ambientales (GIA), Universidad Pontificia Bolivariana, Circular 1 No. 74-50, Medellín 050031, Colombia; juand.martinez@upb.edu.co

*   Correspondence: a.veses@icb.csic.es

Received: 26 March 2020; Accepted: 18 April 2020; Published: 22 April 2020

Abstract: This paper reports the performance of a spark ignition engine using gasoline blended with an upgraded bio-oil rich in aromatics and ethanol. This upgraded bio-oil was obtained using a two-step catalytic process. The first step comprised an in-situ catalytic pyrolysis process with CaO in order to obtain a more stable deoxygenated organic fraction, while the second consisted of a catalytic cracking of the vapours released using ZSM-5 zeolites to obtain an aromatics-rich fraction. To facilitate the mixture between bio-oil and gasoline, ethanol was added. The behaviour of a stationary spark ignition engine G12TFH (9600 W) was described in terms of fuel consumption and electrical efficiency. In addition, gaseous emissions and polycyclic aromatic hydrocarbon (PAH) concentrations were determined. Trial tests suggested that it is possible to work with a blend of gasoline, ethanol and bio-oil (90/8/2 vol%, herein named G90E8B2) showing similar fuel consumption than pure gasoline (G100) at the same load. Moreover, combustion could be considered more efficient when small quantities of ethanol and organic bio-oil are simultaneously added. A reduction, not only in the PAH concentrations but also in the carcinogenic equivalent concentrations, was also obtained, decreasing the environmental impact of the exhaust gases. Thus, results show that it is technically feasible to use low blends of aroma-rich bio-oil, ethanol and gasoline in conventional spark ignition engines.

Keywords: drop-in fuels; biomass; bio-oil; pyrolysis; spark engine; gasoline

---

1. Introduction

Biomass pyrolysis produces solid, gas and liquid fractions. The first one, known as biochar, has several uses as a fuel, in soil remediation and even as a precursor of activated carbon, among others [1,2]. Both gases and liquids are also potentially renewable fuels and they can be used directly or mixed with traditional fuels in conventional internal combustion engines [3]. Pyrolysis gas can be easily used in energy generation, either in spark ignition or compression ignition engines. However, combustion parameters such as calorific value, flame velocity and ignition energy are key parameters regarding the feasibility of using such a low grade gaseous fuel [4]. Regarding to the liquid fraction, known as bio-oil, it still presents several challenges to be used directly to replace fossil fuels in conventional internal combustion engines. The presence of solid particles, high values of acidity, water content and viscosity, storage and thermal instability (due mainly to the presence of a wide range of oxygenated compounds [5]), low cetane number and low heating value prevent its direct application [6]. Thus, upgrading processes must be carried out to incorporate this product into existing infrastructure or to be used directly as drop-in fuel.

The use of bio-oils (both crude and upgraded), especially in compression ignition engines, has been reviewed, and the performance and exhaust emissions have already been studied [4,7,8].

In this sense, different experimental campaigns have reported several difficulties associated with the adverse properties of bio-oil. Mainly, these problems are related to the corrosion and clogging of the injectors, short combustion duration and long ignition delay, difficulties in engine start-up, unstable operation, coke deposition in the piston and cylinders, and subsequent engine seizure [4,9]. Similarly, some experiences show a preheating of the bio-oil prior to combustion. Likewise, costly and complex modifications of the engine can lead to improve the combustion conditions of bio-oil. Among others, these changes include special spark ignition systems for start-up, dual fuel systems, corrosion resistant injectors, and even an increase in the compression ratio [10].

Blending bio-oil with traditional fossil fuels is also an alternative method for utilising this renewable fuel in conventional combustion engines with nor or minor modifications. However, bio-oil blending with commercial fuels is not an easy task because bio-oil is not usually miscible with these fuels [11]. Even using emulsifying agents, stability problems often appear in the bio-oil/diesel blends leading to separation of phases [4]. Nevertheless, some experimental tests have been reported in the literature using bio-oil mixtures with biodiesel, ethanol and diesel fuels in compression ignition engines [9,10,12,13]. Van de Beld et al. [9] used a binary blend of ethanol (30 wt%) and bio-oil (70 wt%) in a modified one-cylinder diesel engine. For that, a stainless steel fuel injector was adapted to the engine. Surprisingly, the ethanol/bio-oil blend improved the overall engine performance, while CO emissions decreased and NO_x emissions increased. However, the modification of the engine might be difficult to justify from an economic point of view. Hossain et al. [4] observed that 20 vol% of bio-oil obtained from de-inking sludge blended with 80 vol% biodiesel can be used in a multi-cylinder diesel engine without adding any ignition additives or surfactants. Although some solid deposits were observed inside the fuel filters, no deterioration in engine condition was observed after 3 h of operation. Pradhan et al. [12] found that the bio-oil from Mahua oilseed has high viscosity and low cetane number, and hence is not suitable for direct utilisation in a diesel engine. However, these authors reported that this bio-oil can be used blended in 30 vol% with diesel fuel in a single-cylinder diesel engine. Lee and Kim [14] examined the performance and emission characteristics of a modified diesel engine (dual-injection diesel engine) using a bio-oil-ethanol mixture. The pilot injection system ensured the proper combustion of high cetane number fuels developing those conditions to keep the main injection stable, which contained the bio-oil-ethanol blend. The results showed that although stable engine operation was possible, the engine efficiency should be improved. Additionally, hydrocarbons (HC) and CO emissions should be reduced.

Although most of the studies reported so far have been carried out using diesel-like fuels blended with bio-oil in compression ignition engines, the use of bio-oil in these engines still faces important challenges to reach a commercial level. On the other hand, the use of these bio-oils or any other kind of bio-fuel derived from them in spark ignition engines is rather rare in the literature. Latest advances in bio-oil upgrading have shown that it is possible to obtain an aroma-rich fraction [15,16]. This fraction has some of the main components of gasoline and could be considered a real additive for this fuel. In addition, upgraded bio-oil/ethanol blends could form a stable mixture without using surfactants or additives. The resulting viscosity is lower than that of pure bio-oil, improving not only atomisation conditions but also the storage and handling properties [17]. The application of a final mixed fuel comprised of gasoline, ethanol and upgraded bio-oil rich in aromatic compounds could emerge as a promising alternative to reduce the use of fossil fuels, while keeping the efficiency of the engine. In this line, the work carried out by Pelaez-Samaniego et al. [18] has shown promising results with a mixture of a fraction derived from a bio-oil obtained in the pyrolysis of sugarcane (rich in esters of carboxylic acids) with gasoline in a spark engine. Specifically, their results showed no significant differences in power and specific fuel consumption, concluding that it is technically feasible to use that mixture in a conventional Otto engine.

However, to the best of our knowledge, there have been no studies on the use of a direct drop-in fuel derived from an upgraded bio-oil in a spark ignition engine. Related studies found in the literature are mainly based on the use of the pyrolytic gas fraction [19]. Thus, successful application of upgraded drop-in fuel obtained from biomass pyrolysis in a spark ignition engine could have a positive impact on actual transportation fuel infrastructure. There are several important parameters, such as fuel consumption and electrical efficiency, that should be addressed to ensure technical feasibility. Moreover, exhaust emissions, including polycyclic aromatic hydrocarbons (PAHs) would provide not only useful information about combustion performance, but also a critical analysis of the air quality from the perspective of human health impacts.

In this work, the organic fraction obtained from a two-step catalytic process was blended in low proportions with commercial gasoline and ethanol. The resulting mixture was fed to a generator group (model G12TFH) equipped with Honda GX620 engine with a common-rail injection system. The main characteristics of the fuel were analysed, and important combustion parameters such as fuel consumption, electrical efficiency, gaseous emissions (O_2, CO, CO_2, CH_4, C_2H_6, C_2H_4, C_3H_8, C_3H_6 and NO_x,), and PAH concentrations were also determined.

2. Materials and Methods

2.1. Raw Bio-Oil

Bio-oil used in this work was obtained using a two-step catalytic process. The first step consisted of a catalytic pyrolysis using pine woodchips as feedstock and CaO as a catalyst. This step was carried out in an auger reactor at pilot scale at 450 °C and using N_2 as carrier gas. Detailed information about experimental conditions and results can be found in previous works [20–22]. The second step also comprised a catalytic process, where the partial-deoxygenated organic fraction produced in the first step was heat-treated at 450 °C in a fixed-bed reactor with hierarchical ZSM-5 zeolite. More details about the catalytic cracking process and chemical characterisation of bio-oil can be found in previous research articles [23–27].

2.2. Gasoline and Ethanol

The gasoline used as commercial fuel reference and for blending was acquired from CEPSA Corporation (Spain). This fuel meets all the requirements established in the specifications defined by Royal Decree 1088/2010. It complies with Directive 2009/30/EC and with the European standard CEN EN 228. On the other hand, commercial ethanol (96 vol%) was provided by Scharlab, S.L.

2.3. Spark Ignition Engine and Generator

The experimental tests were carried out in a generator group (model G12TFH) equipped with an air-cooled, 4-Stroke, overhead valve (OHV), 90 L V-twin Honda GX620 engine with common-rail injection system. Specific characteristics of engine and generator can be seen in Tables 1 and 2, respectively.

Table 1. Main specifications of the spark ignition engine.

Engine Type	Air-Cooled, 4-Stroke, OHV, 90 L V-twin
Bore × Stroke	77 × 66 mm
Displacement	614 cm^3
Compression Ratio	8.3:1
Net Power Output	13.5 kW at 3600 rpm
Net Torque	40.6 Nm at 2500 rpm
Number of cylinders	2
Nominal fuel consumption	4.1 L/h
Dimensions (L × W × H)	388 × 457 × 452 mm
Net Weight	42 kg

Table 2. Main specifications of the generator group.

Start Type	Electric, Automatic
Number of phases	3 (380/220 V/V)
Type of generator	Synchronic
Active power	9.6 kW
Total power	12 kVA
Noise level	77 dB
Dimensions (L × W × H)	900 × 600 × 580 mm
Weight	107 kg

2.4. Characterisation of Organic Bio-Oil Fraction

The organic bio-oil fraction used in this work was analysed by determining different physicochemical properties according to different standard methods. Thus, ultimate composition (Carlo Erba EA1108) and calorific value (IKA C-2000) were determined according to UNE-EN ISO 16948:2015 [28] and UNE-EN ISO 18125:2018 [29], respectively. Water content was also determined by Karl-Fischer titration (Crison Titromatic) in accordance with ASTM E203-96 [30], while pH and total acid number (TAN) were determined at room temperature using a Mettler Toledo T50 device coupled with an electrode Inlab Micro and an electrode DGi118-solvant, respectively. Likewise, density (Antor-Paar DMA35N) and viscosity (Brookfield LVDV-E) were also measured following the standard ASTM D445 [31].

The chemical composition of the organic bio-oil fraction was determined by gas chromatography-mass spectrometry (GC/MS) by means of a Varian CP-3800 gas chromatograph (GC) coupled to a Saturn 2200 Ion Trap mass spectrometer (MS). A capillary column, CP-Sil 8 CB, low bleed: 5% phenyl, 95% dimethylpolysiloxane, (60 m, 0.25 mm i.d., film thickness 0.25 μm film thickness) was used. The oven temperature was initially set at 70 °C and then was kept there for 1 min. Then, a final column temperature of 300 °C was set, implementing a heating rate of 4 °C/min. From this point, the temperature was maintained for 21 min. He (BIP$^®$ quality) was used as the carrier gas was at a constant column flow rate of 1 mlN/min. 1 microliter of samples (1:25 wt%, solvent CH_2Cl_2:C_2H_6O (1:1)) was injected with a solvent delay of 7.5 min and a split ratio 25:1. The temperatures of the injector, detector and transfer line were 280 °C, 200 °C and 300 °C. Electron ionisation mode ranging between 35-550 m/z was used by MS, and compounds were determined by integrating the corresponding m/z. Duplicate analyses were carried out for each sample giving results as an average. At this point, it should be mentioned that relative standard deviation (RSD) values were lower than 10% for all components, and these low variations do not affect the discussion and conclusion sections. The percentage of each compound in the bio-oil was calculated by area normalisation, based on dividing the area of each peak between the total area. In this way, the compounds were classified into families according to their nature. The NIST2011 library was used to identify the mass spectra obtained from the GC/MS analyses.

2.5. Gas Emissions and Polycyclic Aromatic Hydrocarbons (PAHs) Trapping System

Gas emissions (O_2, CO, CO_2 and light hydrocarbons (C_xH_y)) were determined by gas chromatography. For this, several gas samples were analysed (every 5 minutes) in a Hewlett Packard series II GC connected to a Thermal Conductivity Detector (TCD). A Molsieve 5 Å column was used to analyse O_2 and CO (isothermal at 60 °C) and a HayeSep Q column to analyse CO_2 and CxHy (isothermal at 90 °C). In addition, a TESTO 350 XL analyser placed in the exhaust pipe of engine was used to carry out on-line characterisation of some of the gaseous emissions released (O_2, CO_2, NO_x and SO_2). It should be pointed out that the emissions always remained at similar values for all tests (RSD values lower than 7%–5% when gas % or ppm were measured).

A flow sampling system was also installed in the exhaust pipe to capture PAH emissions. The trapping system used for this purpose consisted of a quartz filter (47 mm diameter) to capture the PAHs present in the produced particulate matter and two connected steel tube cartridges filled up with XAD-2 resin (1.5 g each one) supported by quartz wool. The adsorption system was maintained at 120 °C, avoiding water condensation. Filters and resins were cleaned-up by Soxhlet extraction with DCM (24 h) before sampling.

PAHs were quantified using the GC/MS system previously described and following the methodology shown in a previous work [32]. Briefly, it consisted of individual Soxhlet extraction of the filter and each resin with the addition of deuterated-PAH surrogate standards containing different PAHs, such as anthracene-d10 (An-d10), acenaphthene-d10 (Ace-d10), benzo(a)anthracene-d12 (BaA-d12), benzo(ghi)-perylene-d12 (BghiP-d12) and benzo(a)pyrene-d12 (BaP-d12). Samples were concentrated by a rotary evaporator and afterwards by a nitrogen stream flow exchanging the solvent to n-hexane and p-terphenyl (p-tph) native was added as recovery standard.

Each compound was quantified by GC-MS/MS that operates at electron impact energy of 70 eV and utilising the multiple reaction monitoring (MRM) mode (S.1, PAH analysis). PAH quantification was conducted using the internal standard method that is in relation with the closest eluting PAH surrogate. Standard calibrations were prepared at different concentrations obtaining six-point calibration curves with correlation coefficients (r^2) higher than 0.99. Injections were performed in the splitless mode using a selective PAH capillary column (0.25 mm inner diameter and 30 m length and with a film thickness of 0.25 μm). Helium BIP® at a constant flow of 1.5 mL/min was used as a carrier gas. The temperature/time program was 70 °C, 1 min until 325 °C with a ramp of 10 °C/min and this temperature was maintained for 13.5 min. The injector, ion trap and transfer line were kept at 280 °C, 200 °C and 300 °C, respectively.

The following PAHs were quantified by GC-MS/MS, according to their elution time: naphthalene (Np, m/z 102), acenaphthene (Ace, m/z 151), acenaphthylene (Acy, m/z 150), fluorene (Fl, m/z 163), fluoranthene (Fth, m/z 200), phenanthrene (Phe, m/z 152), anthracene (An, m/z 152), 2+2/4-methylphenanthrene (2+2/4MePhe, m/z 189), 9-methylphenanthrene (9MePhe, m/z 189), 1-methylphenanthrene (1MePhe, m/z 189), 2,5-/2,7-/4,5-dimethylphenanthrene (DiMePhe, m/z 191), pyrene (Py, m/z 200), benzo(b)fluoranthene (BbF, m/z 250), benzo(a)anthracene (BaA, m/z 226), chrysene (Chry, m/z 226), benzo(k)fluoranthene (BkF, m/z 250), benzo(a)pyrene (BaP, m/z 250), benzo(e)pyrene (BeP, m/z 250), indeno(1,2,3-cd)pyrene (IcdP, m/z 274), dibenzo(a,h)anthracene (DahA, m/z 276), benzo(ghi)perylene (BghiP, m/z 274) and coronene (Cor, m/z 298) [33]. Detailed information about the conditions for the PAH quantification by GC-MS/MS and PAH compounds and their toxic equivalency factors (TEF) was included in the supporting information (Tables S1 and S2).

2.6. Blending Step

The bio-oil organic fraction was first mixed with ethanol, maintaining a proportion of 20/80 in volume. This sample was kept for 1 day at 25 °C, showing no evidence of phase separation. The resulting mixture was then mixed with commercial gasoline (10/90 in volume) achieving a total mixture of 90/8/2 vol% of gasoline/ethanol/bio-oil. From this point, this sample will be referred to as G90E8B2. No higher proportions were able to be tested due to availability of the bio-oil sample. It should be noted that, based on the amount of available biomass for energy generation and, more concisely, based on the amount of bio-oil that can be generated by pyrolysis processes (which is low in comparison with fossil-based commercial fuels market) [34], this selected proportion is considered as representative to assess the influence of the biofuel as a drop-in fuel [4]. Moreover, it should be mentioned that new legislation concerning the promotion of the use of biofuels, as well as other renewable fuels in transport, proposes to reach higher percentages each year. Particularly, the European Directive 2009/28/EC [35], related to the promotion of the use of energy derived from renewable sources, establishes that each Member State shall guarantee that the share of energy derived from renewable sources concerning all types of transport in 2020 is at least equivalent to 10 percent of its final energy consumption in transport. Hence, considering that the ethanol is obtained from renewable sources, this mixture can be considered as a good approach to assess the results. The resulting sample was kept and filtered, ensuring no solid particles presence in the final mixture. Finally, in order to check the stability of the sample, it was stored for 1 day at 25 °C. After that period of time, no phase separation was observed and sample was considered ready to be used. An aliquot of the mixture can be seen in Figure 1.

Figure 1. Final sample of G90E8B2 after filtering.

2.7. Fuel Consumption and Electrical Efficiency

The fuel consumption is an important parameter to assess the performance of a fuel in an internal combustion engine. The assessment of this parameter estimates the potential fuel savings and costs, as well as reductions in CO_2 emissions. Together with efficiency, this parameter gives a general perspective on how the engine works with a particular fuel. In this case, the electrical efficiency was calculated according to Equation (1) [13]:

$$\text{Electrical Efficiency } (\%) = 100 \times \frac{\varnothing_{fuel} \times \rho_{fuel} \times LHV_{fuel}}{3.6 \times P_e} \tag{1}$$

where $\varnothing_{fuel}$ corresponds with fuel consumption (l/h), ρ_{fuel} with the density (kg/L), LHV_{fuel} with the lower heating value of each fuel (MJ/kg) and P_e with the electrical output (kW), being the same for the two tests (4.4 kW).

3. Results and Discussion

3.1. Properties of the Bio-Oil, Ethanol and Gasoline

Table 3 shows the main physicochemical characteristics of the upgraded bio-oil, the gasoline (G100), the ethanol (EtOH) and the resulting mixture of these fuels (G90E8B2). For comparison purposes, the same characterisations from common bio-oils were also included. Generally speaking, common bio-oils produced from biomass pyrolysis are characterised by a great amount of undesired oxygenated compounds. This composition and the remarkable water content (15–30 wt%), entails a polar nature that avoids a complete blending with commercial hydrocarbons. Moreover, the significant content of organic acids, such as acetic or formic acid, confers it a high acidic character (pH values around 2–3). Among other reasons, common pyrolysis oil presents poor ignition properties, complicating its use in common diesel or spark engines. A complete characterization of common pyrolysis oils can be found in extensive reviews [36]. On the other hand, the upgraded bio-oil used in this study shows interesting properties to be considered as potential drop-in fuel. Although there is still oxygen in the bio-oil, its content is much lower than that found in conventional bio-oils obtained from biomass pyrolysis [37,38]. This elemental composition entails a relative high calorific value (32.5 MJ/kg), which is a very important factor for the purpose of this work. In addition, acidic problems related to common bio-oils are expected to be reduced as higher pH and lower TAN values are observed. Also, it is important to highlight that water content is reduced with a final value lower than 5 wt%, keeping density and viscosity at the same values as those of common bio-oils. Regarding the mixture G90E8B2, it was observed that the proportions of ethanol and bio-oil used barely modified the main properties of the final fuel when compared to the initial gasoline sample. However, the applicability of a prospective bio-fuel can only be analysed by means of test engine runs [39]. For this reason, experiments in the engine were performed without any modification in comparison with the G100 experiment.

Table 3. Some properties of common crude bio-oils (adapted from [8]), organic bio-oil fraction used in this work, gasoline, ethanol and G90E8B2 blend.

Fuel	Ultimate Analysis (%.wt)					Calorific Value (MJ/kg)	Water Content (%.wt)	pH–25 °C	TAN (mgKOH/g)	Viscosity (cP)–40 °C	Density (kg/L)–25 °C
	C	O	H	N	S						
Crude bio-oils	30–40	35–55	6–7	0.2–0.5	0.0	16–17	15–30	2–3	60–90	10–15	1.1–1.4
Upgraded bio-oil	75	16.4	8.4	0.2	0.0	32.5	4.7	4.0	21.3	9.4	1.15
G100	87.2	0.0	12.9	0.0	0.01	43.6	<0.1	2.9	<5	0.12	0.75
EtOH	52.2	34.7	13.1	0.0	0.0	29.6	3.8	7.3	<5	1.9	0.78
G90E8B2	84.2	3.1	12.8	0.004	0.009	42.3	0.4	3.2	<5	0.45	0.77

Regarding the chemical composition of common bio-oils, it can be stated that bio-oil is a complex organic mixture of compounds, mainly alcohols, ketones, phenols, aldehydes, ethers, esters, sugars, furans, alkenes, nitrogen and other different oxygenated compounds [40]. This liquid also presents considerable amount of reactive molecules with high molecular weight, that contribute to their instability [41]. Thus, upgrading post-conversion treatments are necessary to produce potential market bio-products [42,43]. Specifically, to facilitate the compatibility of bio-oil with other fuels, upgrading methods are focused on bio-oil deoxygenation and hydrogenation. Our research group has reached important advances in this field, achieving a great reduction of the oxygen content in the final bio-oil, as well as increasing the production of value-added products like aromatics. In this case, after the two-step upgrade process, it has been possible to obtain a significant deoxygenated organic phase rich in light aromatic compounds. These kinds of hydrocarbons make up a large part of gasoline, containing up to 35 vol%. Other hydrocarbons are olefins (18 vol%) and benzene (1 vol%) according to technical data files of CEPSA corporation [44]. It should be pointed out that the aromatic fraction of the upgraded bio-oil is mainly composed of C_6–C_8 aromatics (40–60 %). These components can be considered to be benzene, toluene and xylene (BTX) hydrocarbons, which are very valuable aromatic compounds. Thus, due to its properties and chemical composition, this liquid has been considered a potential renewable hydrocarbon biofuel able to be used as infrastructure-compatible fuel, i.e., as drop-in fuel. Detailed information about this process can be consulted in previous works [20,21,23,24].

3.2. Performance of Engine Tests

Experiments were conducted showing no operational problems and no difficulties in the start-up of the engine, which is one of the common problems when bio-oil is used as fuel in internal combustion engines [4,7]. It should be mentioned that, unfortunately, due to the availability of the bio-oil sample, it was only possible to check one mixture at one operational condition. However, for these conditions, the test was carried out over approximately one hour, showing no symptoms of malfunction. Hence, we consider that these results could be a very interesting starting point for future research, encouraging its application at a higher scale. Indeed, our research group has achieved recent advances in the field of drop-in fuels at higher scales [45], which could also have a great potential for this application. Interestingly, there is no remarkable differences in the specific fuel consumption when the G100 (0.74 g/kWh) and G90E8B2 (0.77 g/kWh) are used at the same load. This suggests that fuel mixture G90E8B2 could maintain the same power consumption as that of commercial gasoline, the small differences being related to the slightly lower calorific value of the mixed fuel [46]. These results are in line with those carried out by Pelaez-Samaniego et al. [18]. In that work, no significant differences in the fuel consumption were found when a bio-oil/gasoline blend (90/10 vol%) was used in a spark ignition engine. Likewise, the presence of ethanol in bio-oil has the advantage of improving viscosity and combustion conditions that seem to be favoured because of the higher octane number and higher flame speed compared to gasoline [47,48]. In addition, the electrical efficiency of the different fuels at the same load was found to be 9.72% and 9.99% for G100 and G90E8B2, respectively. Even so, it can be highlighted that overall efficiencies were very similar for both fuels but slightly superior for G90E8B2 sample.

3.3. Exhaust Gaseous Emissions

Figure 2 presents the gas emissions obtained from the GC analysis for O_2, CO and CO_2, for both fuels (G100 and G90E8B2) under steady state conditions. As it can be observed, oxygen amount in the exhaust gas is lower when the fuel mixture G90E8B2 is fed (16.5 g/kWh), as compared to that found for the G100 (42.0 g/kWh). This is directly related to the increase of CO_2 emissions, which is higher for G90E8B2 (107.8 g/kWh) than for G100 (50.4 g/kWh). In addition, CO emissions were slightly lower for G90E8B2 (9.03 g/kWh) than for G100 (9.63 g/kWh). CO is a product of incomplete combustion, but it also depends on both the temperature in the cylinder and the oxygen content present in the fuel. Although it was not included for brevity and clarity, these tendencies of O_2, CO and CO_2 emissions at steady state could be verified by the results obtained using TESTO XL 350. O_2, CO and CO_2 emissions suggest a better combustion performance of the G90E8B2, mainly due to the fact that the oxygen content provided by both ethanol and bio-oil is higher. The availability of oxygen in G90E8B2 helps to reduce CO at expenses of CO_2. Similar results have been reported in the literature. For instance, Deng et al. [49] observed in a four-cylinder spark ignition engine operated at different engine speeds that increasing ethanol concentration in a blend with gasoline lead to lower CO emissions than pure gasoline. Sakthivel et al. [50] also reported a decrease in the CO emissions with increasing ethanol content as compared to pure gasoline using a single-cylinder spark-ignition engine.

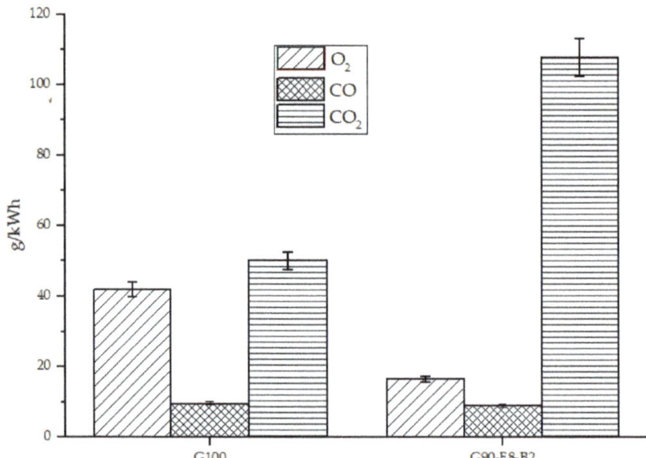

Figure 2. O_2, CO and CO_2 emissions at steady state obtained from the gas chromatograph (GC) analysis.

Another parameter that can provide some useful information about combustion efficiency is the amount of hydrocarbons (HC) emitted. Although differences are not very remarkable, the results shown in Figure 3 confirm that the total HC emissions were slightly higher when commercial gasoline was used instead of G90E8B2.

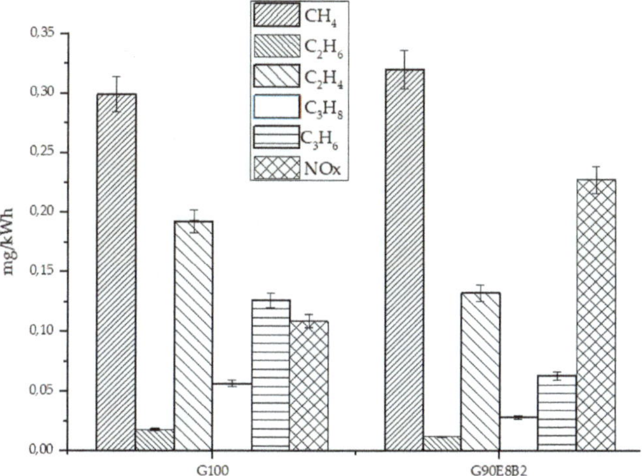

Figure 3. Hydrocarbon and NOx emissions at steady state obtained from the GC analysis.

NO$_x$ emissions are another key parameter in assessing the environmental impact of any fuel used in an internal combustion engine. It is well known that thermal NO$_x$ generation mainly depends on temperature in the cylinder, the oxygen concentration in the combustion reaction, and the residence time. Ethanol has a higher latent heat and a lower energy density than gasoline, and hence a lower temperature is expected. However, as can be seen in Figure 3, NO$_x$ emissions (either NO and NO$_2$) were higher when G90E8B2 (0.22 mg/kWh) was used, as compared to G-100 (0.11 mg/kWh). Deng et al. [49] also reported higher NO$_x$ emissions when the proportions of ethanol increased in a gasoline blend. This could be explained by the slightly higher nitrogen content accounted for the upgraded bio-oil (see Table 3). However, it cannot be completely ruled out that the increase in NO$_x$ emissions are directly related to the temperature and oxygen content in the combustion chamber, or even due to the higher water content of the mixture [12]. This fact could be also explained by the different value of the viscosity found for the mixture that could entail a different behaviour in the atomisation of the fuel in the cylinder [9]. It is also worth mentioning that SO$_2$ was also evidenced, but only was significant in the first minutes of the experiment. While 3.44 mg/kWh were measured when G100 is tested, 0.68 mg/kWh were evidenced for the mixture G90E8B2. Although these can be considered small quantities, the fact of their incorporation in bio-oil can reduce environmental issues related to this component. This could be explained by the negligent sulphur content of the upgraded bio-oil fraction. If ethanol comes from renewable sources, its carbon footprint will be significantly reduced.

3.4. PAH Concentrations

Polycyclic aromatic hydrocarbons (PAH) can be produced because of the incomplete combustion of organic compounds. In any energy generation process, the identification and quantification of these compounds is very important owing to their carcinogenic and mutagenic effects with harmful impacts on human health [51]. This is one of the main reasons why 16 PAHs were recognised by the United States Environmental Protection Agency (US EPA) as a strong carcinogenic teratogenic PAH [52].

PAH concentrations for the two fuels (G100 and G90E8B2) were determined according to the sampling system; a Teflon filter and two simultaneous XAD®-2 resins (Figure 4). Each sample was injected by duplicate and relative standard deviation lower than 10% was obtained for each PAH, in each individual trap system. It was observed that the total PAH concentration was remarkably lower when using the G90E8B2 mixture (4164 ng/m^3) than the one obtained working with G100 (6914 ng/m^3). Moreover, this reduction was mainly appreciated in the resins system (gas phase PAHs). It is also important to highlight the distribution of PAHs according to their volatility as a function of the trapping system. In this regard, the resin was able to trap most of the volatile PAHs, whereas the Teflon filter retained a wider range of PAHs, including heavier PAH-like Cor. According to these results, it seems that the introduction of oxygen into the fuel provided by the ethanol and the upgraded bio-oil improved the efficiency of the combustion process, producing a smaller amount of PAHs, in particular the most volatile ones, which were mainly retained in both resins. These results were in accordance with the gas emission profile showed in the previous section, since oxygen values were lower for the G90E8B2. Other authors [53,54] already observed that oxygenated fuels decreased the soot precursor species during the fuel-rich premixed ignition and promoted the oxidation of the already-formed soot in a diffusion flame. Because certain unsaturated hydrocarbons, such as some PAH and acetylene, are the most likely precursors of soot particles [14], carbon atoms bonded to oxygen atoms in oxygenated fuels do not influence particulate matter formation, and this could explain the low PAH concentration from G90E8B2 combustion.

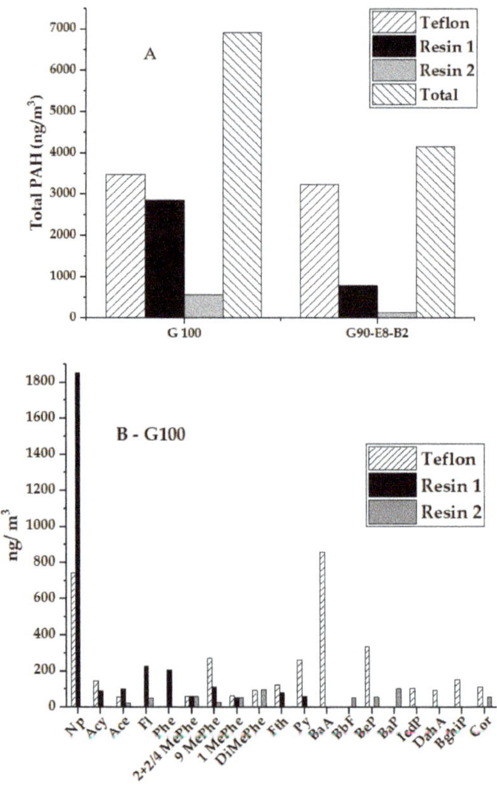

Figure 4. *Cont.*

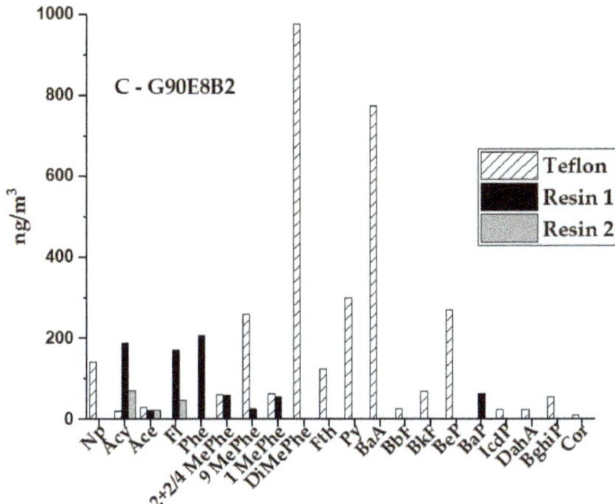

Figure 4. (**A**): Polycyclic aromatic hydrocarbon (PAH) concentrations for each fuel as a function of the trap system. (**B,C**): Distribution of PAH according to their volatility as a function of the trapping system.

This could be also explained according to the nature of the fuel burnt. Authors like Yinhui et al. [55] found that higher concentration of aromatics in gasoline resulted in much higher particulate matter (PM) and PAH emissions. Reducing aromatic compounds content in gasoline by the addition of other fuel component, such as bio-oil/ethanol blend, could result as an important approach to reduce primary particulate emissions of gasoline vehicles and, in consequence, improve air quality [55]. Results found in this work corroborated the lower PAH concentration in G90E8B2 versus the G100 with the substitution of a fraction of gasoline by an ethanol/upgraded bio-oil blend.

As mentioned above, PAH are listed as priority pollutants by the US EPA and they are released to the atmosphere after combustion processes. An interesting parameter to consider, in addition to PAH concentrations, is related to their carcinogenic character. One of the most carcinogenic PAHs is benzo(a)pyrene. In fact, this PAH is regulated by European directives (Directive 2004/107/EC [56]) setting a target value for PAHs in terms of concentration of BaP of 1.0 ng/m^3 in the air as annual mean value measured in the particulate matter equal to or less than 10 µm (PM10) in aerodynamic diameter. In this way, an assessment of this character as a function of the different fuels based on toxic equivalency quotient (TEQ) relative to BaP (BaP-TEQ) was calculated. This behaviour was calculated by considering the toxic equivalency factors (TEF) of Nisbet and LaGoy (1992) [57] (Table S2) and multiplying the TEF by the concentration of each individual PAH in each sample. Results obtained for the two experiments were expressed as a function of the trapping system as BaP-TEQ (Figure 5A). It was observed that the fuel composed of the upgraded bio-oil presented a lower carcinogenicity factor than using pure gasoline. Despite some PAHs like DahA did not show a high contribution to the total PAHs, its contribution to the BaP-TEQ was remarkable, whereas other compounds like Np with high contribution in the G100 combustion were negligible in the BaP-TEQ concentration (Figure 5B). Independently of the fuel burnt, BaA, BaP and IcdP were the PAH with the maximum contribution to the BaP-TEQ. These results reflected the importance in studying PAH emissions, not only PAH concentrations but also the nature and the carcinogenic character of the individual PAH.

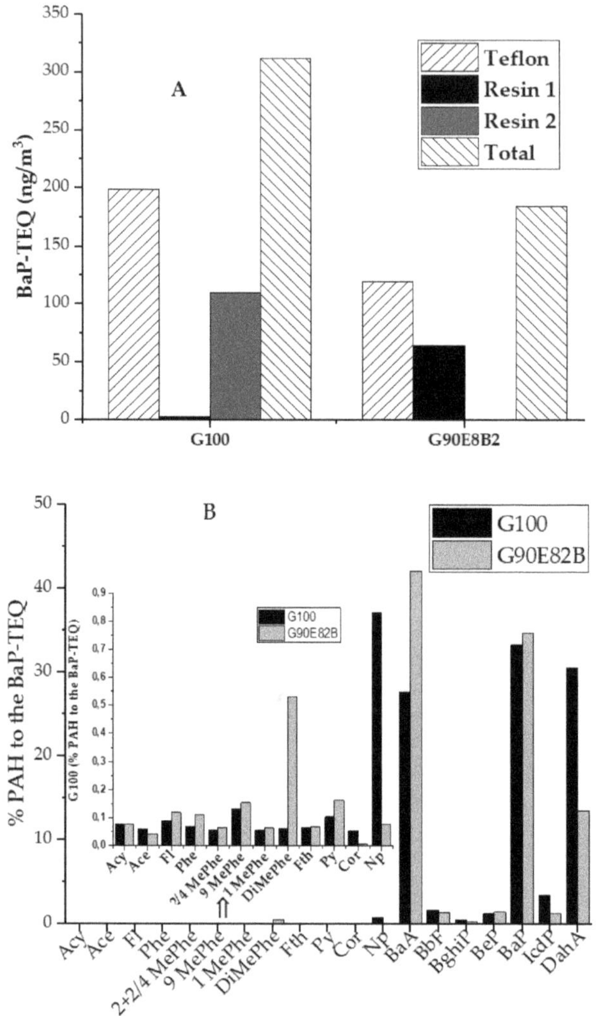

Figure 5. (**A**): Benzo(a)pyrene (BaP)-TEQ (ng/m^3) as a function of the fuel for each individual trap systems. (**B**): percentage of contribution of each individual PAH to the BaP-TEQ for each fuel.

The promising results achieved in this work could offer the first insights on the potential application of upgraded bio-oil fraction as a drop-in fuel in current commercial engines. Extended duration tests and improving bio-oil properties, jointly with the study of different proportions in the final product, are the next steps to consider this application at higher scales.

4. Conclusions

In this work, an upgraded bio-oil/ethanol blend has been successfully tested as a drop-in fuel in a spark ignition engine. This renewable hydrocarbon biofuel can therefore be considered as an infrastructure-compatible fuel. Ethanol was added to facilitate the mixture and stability. The results revealed that it is possible to work with a blend 90/8/2 vol% of gasoline, ethanol and upgraded bio-oil fraction evidencing similar specific fuel consumption and electrical efficiency. Moreover, exhaust emissions suggest that combustion can be considered more completed. Likewise, this blend produces lower PAH concentrations than those with pure gasoline, mainly in the gas phase. Moreover, these emissions are less carcinogenic, which strongly improves the air quality from human health impact. Thus, the results showed that it is technically feasible to use low blends of aromatics-rich bio-oil and ethanol with gasoline in conventional spark engines.

Supplementary Materials: The following are available online at http://www.mdpi.com/1996-1073/13/8/2089/s1, S.1 PAH analysis, Table S1: Composition conditions for PAH quantification by GC-MS-MS, Table S2: PAH compounds and their toxic equivalency factors (TEF).

Author Contributions: Conceptualization, T.G. and R.M.; methodology, A.V.; validation, A.V. and J.D.M.; formal analysis; A.V., J.D.M. and M.S.C.; investigation, A.V., J.D.M., M.S.C.; resources, T.G. and R.M.; data curation, A.V., J.D.M. and M.S.C.; writing—original draft preparation, A.V., J.D.M. and M.S.C.; writing—review and editing, T.G. and R.M.; supervision, T.G. and M.S.C.; project administration, T.G. and M.S.C.; funding acquisition, T.G. and R.M. All authors have read and agreed to the published version of the manuscript.

Funding: This research was funded by MINECO and FEDER (Project ENE2015-68320-R) and the Regional Government of Aragon (DGA) under the research groups support programme.

Conflicts of Interest: The authors declare no conflict of interest.

References

1. Yang, S.I.; Wu, M.S.; Wu, C.Y. Application of biomass fast pyrolysis part I: Pyrolysis characteristics and products. *Energy* **2014**, *66*, 162–171. [CrossRef]
2. Peters, J.F.; Iribarren, D.; Dufour, J. Biomass Pyrolysis for Biochar or Energy Applications? A Life Cycle Assessment. *Environ. Sci. Technol.* **2015**, *49*, 5195–5202. [CrossRef] [PubMed]
3. Mortensen, P.M.; Grunwaldt, J.D.; Jensen, P.A.; Knudsen, K.G.; Jensen, A.D. A review of catalytic upgrading of bio-oil to engine fuels. *Appl. Catal. A Gen.* **2011**, *407*, 1–19. [CrossRef]
4. Hossain, A.K.; Davies, P.A. Pyrolysis liquids and gases as alternative fuels in internal combustion engines—A review. *Renew. Sustain. Energy Rev.* **2013**, *21*, 165–189. [CrossRef]
5. Cheng, F.; Bayat, H.; Jena, U.; Brewer, C.E. Impact of feedstock composition on pyrolysis of low-cost, protein- and lignin-rich biomass: A review. *J. Anal. Appl. Pyrolysis* **2020**, *147*, 104780. [CrossRef]
6. Fermoso, J.; Pizarro, P.; Coronado, J.M.; Serrano, D.P. Advanced biofuels production by upgrading of pyrolysis bio-oil. *Wiley Interdiscip. Rev. Energy Environ.* **2017**, *6*, e245. [CrossRef]
7. Krutof, A.; Hawboldt, K. Blends of pyrolysis oil, petroleum, and other bio-based fuels: A review. *Renew. Sustain. Energy Rev.* **2016**, *59*, 406–419. [CrossRef]
8. No, S.-Y. Application of bio-oils from lignocellulosic biomass to transportation, heat and power generation—A review. *Renew. Sustain. Energy Rev.* **2014**, *40*, 1108–1125. [CrossRef]
9. Van de Beld, B.; Holle, E.; Florijn, J. The use of pyrolysis oil and pyrolysis oil derived fuels in diesel engines for CHP applications. *Appl. Energy* **2013**, *102*, 190–197. [CrossRef]
10. Hossain, A.K.; Ouadi, M.; Siddiqui, S.U.; Yang, Y.; Brammer, J.; Hornung, A.; Kay, M.; Davies, P.A. Experimental investigation of performance, emission and combustion characteristics of an indirect injection multi-cylinder CI engine fuelled by blends of de-inking sludge pyrolysis oil with biodiesel. *Fuel* **2013**, *105*, 135–142. [CrossRef]
11. Hansen, S.; Mirkouei, A.; Diaz, L.A. A comprehensive state-of-technology review for upgrading bio-oil to renewable or blended hydrocarbon fuels. *Renew. Sustain. Energy Rev.* **2020**, *118*, 109548. [CrossRef]
12. Pradhan, D.; Bendu, H.; Singh, R.K.; Murugan, S. Mahua seed pyrolysis oil blends as an alternative fuel for light-duty diesel engines. *Energy* **2017**, *118*, 600–612. [CrossRef]
13. Van de Beld, B.; Holle, E.; Florijn, J. The use of a fast pyrolysis oil—Ethanol blend in diesel engines for chp applications. *Biomass Bioenergy* **2018**, *110*, 114–122. [CrossRef]
14. Lee, S.; Kim, T.Y. Feasibility study of using wood pyrolysis oil–ethanol blended fuel with diesel pilot injection in a diesel engine. *Fuel* **2015**, *162*, 65–73. [CrossRef]
15. Rezaei, P.S.; Shafaghat, H.; Daud, W.M.A.W. Production of green aromatics and olefins by catalytic cracking of oxygenate compounds derived from biomass pyrolysis: A review. *Appl. Catal. A Gen.* **2014**, *469*, 490–511. [CrossRef]
16. Rahman, M.M.; Liu, R.; Cai, J. Catalytic fast pyrolysis of biomass over zeolites for high quality bio-oil—A review. *Fuel Process. Technol.* **2018**, *180*, 32–46. [CrossRef]

17. Nguyen, D.; Honnery, D. Combustion of bio-oil ethanol blends at elevated pressure. *Fuel* **2008**, *87*, 232–243. [CrossRef]

18. Pelaez-Samaniego, M.R.; Mesa-Pérez, J.; Cortez, L.A.B.; Rocha, J.D.; Sanchez, C.G.; Marín, H. Use of blends of gasoline with biomass pyrolysis-oil derived fractions as fuels in an Otto engine. *Energy Sustain. Dev.* **2011**, *15*, 376–381. [CrossRef]

19. Dupuis, D.P.; Grim, R.G.; Nelson, E.; Tan, E.C.D.; Ruddy, D.A.; Hernandez, S.; Westover, T.; Hensley, J.E.; Carpenter, D. High-Octane Gasoline from Biomass: Experimental, Economic, and Environmental Assessment. *Appl. Energy* **2019**, *241*, 25–33. [CrossRef]

20. Veses, A.; Aznar, M.; Martínez, I.; Martínez, J.D.; López, J.M.; Navarro, M.V.; Callén, M.S.; Murillo, R.; García, T. Catalytic pyrolysis of wood biomass in an auger reactor using calcium-based catalysts. *Bioresour. Technol.* **2014**, *162*, 250–258. [CrossRef]

21. Veses, A.; Aznar, M.; Callén, M.S.; Murillo, R.; García, T. An integrated process for the production of lignocellulosic biomass pyrolysis oils using calcined limestone as a heat carrier with catalytic properties. *Fuel* **2016**, *181*, 430–437. [CrossRef]

22. Veses, A.; Aznar, M.; López, J.M.; Callén, M.S.; Murillo, R.; García, T. Production of upgraded bio-oils by biomass catalytic pyrolysis in an auger reactor using low cost materials. *Fuel* **2015**, *141*, 17–22. [CrossRef]

23. Puértolas, B.; Veses, A.; Callén, M.S.; Mitchell, S.; García, T.; Pérez-Ramírez, J. Porosity-Acidity Interplay in Hierarchical ZSM-5 Zeolites for Pyrolysis Oil Valorization to Aromatics. *ChemSusChem* **2015**, *8*, 3283–3293. [CrossRef]

24. Veses, A.; Puértolas, B.; López, J.M.; Callén, M.S.; Solsona, B.; García, T. Promoting Deoxygenation of Bio-Oil by Metal-Loaded Hierarchical ZSM-5 Zeolites. *ACS Sustain. Chem. Eng.* **2016**, *4*, 1653–1660. [CrossRef]

25. Veses, A.; Puértolas, B.; Callén, M.S.; García, T. Catalytic upgrading of biomass derived pyrolysis vapors over metal-loaded ZSM-5 zeolites: Effect of different metal cations on the bio-oil final properties. *Microporous Mesoporous Mater.* **2015**, *209*, 189–196. [CrossRef]

26. García, T.; Veses, A.; López, J.M.; Puértolas, B.; Pérez-Ramírez, J.; Callén, M.S. Determining Bio-Oil Composition via Chemometric Tools Based on Infrared Spectroscopy. *ACS Sustain. Chem. Eng.* **2017**, *5*, 8710–8719. [CrossRef]

27. Veses, A.; López, J.M.; García, T.; Callén, M.S. Prediction of elemental composition, water content and heating value of upgraded biofuel from the catalytic cracking of pyrolysis bio-oil vapors by infrared spectroscopy and partial least square regression models. *J. Anal. Appl. Pyrolysis* **2018**, *132*, 102–110. [CrossRef]

28. AENOR. *UNE-EN ISO 16948:2015. Solid Biofuels-Determination of Total Content of Carbon, Hydrogen and Nitrogen*; AENOR: Madrid, Spain, 2015.

29. AENOR. *UNE-EN ISO 18125:2018. Solid Biofuels-Determination of Calorific Value*; AENOR: Madrid, Spain, 2018.

30. ASTM. *ASTM E203-96. Standard Test Method for Water Using Volumetric Karl Fischer Titration*; ASTM International: West Conshohocken, PA, USA, 2016.

31. ASTM. *ASTM D445-19. Standard Test Method for Kinematic Viscosity of Transparent and Opaque Liquids (and Calculation of Dynamic Viscosity)*; ASTM International: West Conshohocken, PA, USA, 2019.

32. Callén, M.S.; de la Cruz, M.T.; López, J.M.; Murillo, R.; Navarro, M.V.; Mastral, A.M. Some inferences on the mechanism of atmospheric gas/particle partitioning of polycyclic aromatic hydrocarbons (PAH) at Zaragoza (Spain). *Chemosphere* **2008**, *73*, 1357–1365. [CrossRef]

33. Callén, M.S.; Iturmendi, A.; López, J.M. Source apportionment of atmospheric PM2.5-bound polycyclic aromatic hydrocarbons by a PMF receptor model. Assessment of potential risk for human health. *Environ. Pollut.* **2014**, *195*, 167–177. [CrossRef]

34. Patel, M.; Zhang, X.; Kumar, A. Techno-economic and life cycle assessment on lignocellulosic biomass thermochemical conversion technologies: A review. *Renew. Sustain. Energy Rev.* **2016**, *53*, 1486–1499. [CrossRef]

35. European Commission. Directive 2009/28/EC of the European Parliament and of the Council of 23 April 2009 on the promotion of the use of energy from renewable sources. *Off. J. Eur. Union* **2009**, *L140*, 16–62.

36. Oasmaa, A.; Peacocke, C. *A Guide to Physical Property Characterisation of Biomass-Derived Fast Pyrolysis Liquids*; VTT Publications; Technical Research Centre of Finland: Espoo, Finland, 2010; pp. 2–65.

37. Chiaramonti, D.; Oasmaa, A.; Solantausta, Y. Power generation using fast pyrolysis liquids from biomass. *Renew. Sustain. Energy Rev.* **2007**, *11*, 1056–1086. [CrossRef]

38. Meier, D.; Van De Beld, B.; Bridgwater, A.V.; Elliott, D.C.; Oasmaa, A.; Preto, F. State-of-the-art of fast pyrolysis in IEA bioenergy member countries. *Renew. Sustain. Energy Rev.* **2013**, *20*, 619–641. [CrossRef]
39. Harnisch, F.; Blei, I.; dos Santos, T.R.; Möller, M.; Nilges, P.; Eilts, P.; Schröder, U. From the test-tube to the test-engine: Assessing the suitability of prospective liquid biofuel compounds. *RSC Adv.* **2013**, *3*, 9594–9605. [CrossRef]
40. Hu, X.; Gholizadeh, M. Biomass pyrolysis: A review of the process development and challenges from initial researches up to the commercialisation stage. *J. Energy Chem.* **2019**, *39*, 109–143. [CrossRef]
41. Garcìa-Pérez, M.; Chaala, A.; Pakdel, H.; Kretschmer, D.; Rodrigue, D.; Roy, C. Multiphase Structure of Bio-oils. *Energy Fuels* **2006**, *20*, 364–375. [CrossRef]
42. Makarfi Isa, Y.; Ganda, E.T. Bio-oil as a potential source of petroleum range fuels. *Renew. Sustain. Energy Rev.* **2018**, *81*, 69–75. [CrossRef]
43. Oh, Y.-K.; Hwang, K.-R.; Kim, C.; Kim, J.R.; Lee, J.-S. Recent developments and key barriers to advanced biofuels: A short review. *Bioresour. Technol.* **2018**, *257*, 320–333. [CrossRef]
44. Cepsa Global Energy Company Home Page. Available online: http://www.cepsa.com (accessed on 24 March 2020).
45. Sanahuja-Parejo, O.; Veses, A.; López, J.M.; Murillo, R.; Callén, M.S.; García, T. Ca-based Catalysts for the Production of High-Quality Bio-Oils from the Catalytic Co-Pyrolysis of Grape Seeds and Waste Tyres. *Catalysts* **2019**, *9*, 992. [CrossRef]
46. Martínez, J.D.; Rodríguez-Fernández, J.; Sánchez-Valdepeñas, J.; Murillo, R.; García, T. Performance and emissions of an automotive diesel engine using a tire pyrolysis liquid blend. *Fuel* **2014**, *115*, 490–499. [CrossRef]
47. Bergthorson, J.M.; Thomson, M.J. A review of the combustion and emissions properties of advanced transportation biofuels and their impact on existing and future engines. *Renew. Sustain. Energy Rev.* **2015**, *42*, 1393–1417. [CrossRef]
48. Zhang, Z.; Wang, T.; Jia, M.; Wei, Q.; Meng, X.; Shu, G. Combustion and particle number emissions of a direct injection spark ignition engine operating on ethanol/gasoline and n-butanol/gasoline blends with exhaust gas recirculation. *Fuel* **2014**, *130*, 177–188. [CrossRef]
49. Deng, X.; Chen, Z.; Wang, X.; Zhen, H.; Xie, R. Exhaust noise, performance and emission characteristics of spark ignition engine fuelled with pure gasoline and hydrous ethanol gasoline blends. *Case Stud. Therm. Eng.* **2018**, *12*, 55–63. [CrossRef]
50. Sakthivel, P.; Subramanian, K.A.; Mathai, R. Experimental study on unregulated emission characteristics of a two-wheeler with ethanol-gasoline blends (E0 to E50). *Fuel* **2020**, *262*, 116504. [CrossRef]
51. World Health Organization. Environmental health criteria 229: Nitrogenated polycyclic aromatic hydrocarbons. In *WHO Report Number. International Program on Chemical Safety*; WHO, Ed.; World Health Organization: Geneva, Switzerland, 2004.
52. [ATSDR] Agency for Toxic Substances and Disease Registry. *Toxicological Profile for Polycyclic Aromatic Hydrocarbons (PAHs) (Update)*; US Department of Health and Human Services: Atlanta, GA, USA, 1995.
53. Song, J.; Cheenkachorn, K.; Wang, J.; Perez, J.; Boehman, A.L.; Young, P.J.; Waller, F.J. Effect of Oxygenated Fuel on Combustion and Emissions in a Light-Duty Turbo Diesel Engine. *Energy Fuels* **2002**, *16*, 294–301. [CrossRef]
54. Westbrook, C.K.; Pitz, W.J.; Curran, H.J. Chemical Kinetic Modeling Study of the Effects of Oxygenated Hydrocarbons on Soot Emissions from Diesel Engines. *J. Phys. Chem. A* **2006**, *110*, 6912–6922. [CrossRef]
55. Yinhui, W.; Rong, Z.; Yanhong, Q.; Jianfei, P.; Mengren, L.; Jianrong, L.; Yusheng, W.; Min, H.; Shijin, S. The impact of fuel compositions on the particulate emissions of direct injection gasoline engine. *Fuel* **2016**, *166*, 543–552. [CrossRef]
56. European Commission. Directive 2004/107/EC of the European Parliament and of the Council of 15 December 2004 relating to arsenic, cadmium, mercury, nickel and polycyclic aromatic hydrocarbons in ambient air. *Off. J. Eur. Union* **2004**, *L23*, 3–16.
57. Nisbet, I.C.T.; LaGoy, P.K. Toxic equivalency factors (TEFs) for polycyclic aromatic hydrocarbons (PAHs). *Regul. Toxicol. Pharmacol.* **1992**, *16*, 290–300. [CrossRef]

Article

Nanostructured Carbon Material Effect on the Synthesis of Carbon-Supported Molybdenum Carbide Catalysts for Guaiacol Hydrodeoxygenation

Elba Ochoa, Daniel Torres, José Luis Pinilla * and Isabel Suelves

Instituto de Carboquímica, CSIC, C/Miguel Luesma Castán 4, 50018 Zaragoza, Spain; eochoa@icb.csic.es (E.O.); dtorres@icb.csic.es (D.T.); isuelves@icb.csic.es (I.S.)
* Correspondence: jlpinilla@icb.csic.es; Tel.: +34-976-733-977

Received: 15 January 2020; Accepted: 3 March 2020; Published: 5 March 2020

Abstract: The impact of using different nanostructured carbon materials (carbon nanofibers, carbon nanotubes, graphene oxide and activated carbon) as a support for Mo_2C-based catalysts on the hydrodeoxygenation (HDO) of guaiacol was studied. To optimise the catalyst preparation by carbothermal hydrogen reduction (CHR), a thermogravimetric study was conducted to select the optimum CHR temperature for each carbon material, considering both the crystal size of the resulting β-Mo_2C particles and the extent of the support gasification. Subsequently, catalysts were prepared in a fixed bed reactor at the optimum temperature. Catalyst characterization evidenced the differences in the catalyst morphology as compared to those prepared in the thermogravimetric study. The HDO results demonstrated that the carbon nanofiber-based catalyst was the one with the best catalytic performance. This behaviour was attributed to the high thermal stability of this support, which prevented its gasification and promoted a good evolution of the crystal size of Mo species. This catalyst exhibited well-dispersed β-Mo_2C nanoparticles of ca. 11 nm. On the contrary, the other supports suffered from severe gasification (60–70% wt. loss), which resulted in poorer HDO efficiency catalysts regardless of the β-Mo_2C crystal size. This exhibited the importance of the carbon support stability in Mo_2C-based catalysts prepared by CHR.

Keywords: Mo_2C catalysts; nanostructured carbon materials; hydrodeoxygenation of guaiacol; carbothermal hydrogen reduction

1. Introduction

Since the development in 1985 by Boudart et al. of high specific surface area carbides and nitrides by temperature-programmed reduction [1], several authors have synthetised Mo_2C catalysts by this method using hydrocarbon/hydrogen mixtures [2–8]. Alternatively, for carbon-supported Mo_2C catalysts, the carbothermal hydrogen reduction (CHR) method may transform Mo oxides into hexagonal-close-packed carbides (β-Mo_2C) at relatively moderate temperatures (<800 °C) using pure hydrogen [9–14]. CHR has attracted scientific attention due to the fact that it might avoid carbide contamination by polymeric carbon deposition on the active sites when hydrocarbons are used as a carbon source [15]. Besides, the use of milder temperature conditions may mitigate the low specific surface area obtained due to the partial destruction of the support [16]. Nevertheless, several parameters are involved in the resulting crystallographic and morphological characteristics of the Mo_2C phase, such as carburization temperature, heating rate, carburization time, Mo precursor, Mo content and the nature of the carbon support [11,12,14,17–19].

The carbon support, besides being the carbon source in CHR, is regarded as one of the best choices as a support in biomass-based liquid phase reactions [15,20–24], with the high efficiency in oxygen removal minimizing the H_2 consumption [25]. Thus, these materials have been widely studied both in the CHR and as catalyst support. Mordenti et al. studied Mo_2C formation from activated carbon (AC) in CHR at 700 °C, showing complete carburization which led to Mo_2C crystals of 14 nm on average. In addition, the authors observed an increase of the metal loading and a depletion of the specific surface area after CHR due to gasification [9]. Pielaszek et al. compared the use of two different ACs (NC100 and carbon monolith from expanded graphite) in the formation of Mo_2C by CHR at temperatures up to 700 °C [26]: for the former support, at 700 °C, the authors observed the formation of hexagonal close packing Mo_2C crystals smaller than 10 nm; for the latter, this crystal formation took place at lower temperatures (450 °C). The monolith also showed a Mo metal signal in some cases. Additionally, the authors explored different carburization times (1 h and 108 h) for this support at relatively low temperatures 450–600 °C. Longer times led to a slight increase of the crystal size in contrast to the relative high increase obtained in the Mo metal phase. Liang et al. [10] used a high surface area carbon obtained from petroleum coke as the support of a Mo catalyst with 10% Mo content by CHR. They obtained 6 nm Mo_2C crystals at 700 °C; this was accompanied with a surface area depletion from 3234 m^2/g to 2341 m^2/g [10]. Wang et al. [12] used a similar petcoke with 10%Mo; in contrast, they determined 750 °C to be the optimum carburization temperature to obtain a mesoporous catalyst without suffering Mo_2C sintering and support collapse. In turn, they extended the carburization time for 90 min, provoking an enlargement of the Mo_2C crystal and an increase on the mesoporous volume. This resulted in a higher catalytic activity in 4-methyl phenol hydrodeoxygenation (HDO) [12]. In order to determine the carburization temperature effect in CHR using an activated charcoal, Wang et al. prepared Mo_2C/AC catalysts varying the CHR temperature from 600 °C to 800 °C. They concluded that an increase in temperature led to a higher crystallization degree and less oxygen in the intermediate MoO_xC_y species commonly obtained in the carburization process. However, an overpass at 700 °C resulted in a metallic agglomeration, which diminished the catalytic activity in the HDO of 4 methyl phenol [11]. He et al. studied the effect of Mo loading (10 wt.%, 20 wt.%, 30 wt.%, 50 wt.% and 80 wt.%) on raw AC activated with HNO_3 on the CHR at 700 °C. Higher Mo loading resulted in a decrease of the BET surface area and total volume pore due to the partial destruction of the support and in turn resulted in larger Mo_2C crystals and agglomerations [27]. They tested these catalysts in the vanillin HDO, concluding that the optimum Mo loading was 20 wt.% due to a good balance between the Mo content and the absence of particle agglomeration [27].

Nanofilamentous supports such as carbon nanofibers (CNF) and nanotubes (CNT) were also studied in CHR. For CNT, Frank et al. evaluated the CHR conditions (gas flow, heating rate and Mo loading), and they noticed that the defects of the CNT structure were crucial in the Mo_xC crystal formation, which led to a highly active catalyst in the steam reforming of methanol (SRM) [28]. Subsequently, aiming to enhance the Mo_2C/CNT properties in this reaction, they modified the CNT surface with heteroatoms (O, N and S). The performed functionalization affected the final nature of the Mo_2C nanoparticles and, in particular, N-doped CNT increased the Mo dispersion and the catalytic activity [29]. For the same purpose, Barthos et al. transformed the Mo precursor (ammonium heptamolybdate) into MoO_3 prior to CHR and used the resulting Mo_2C/CNT as a catalyst for SRM. In the preparation, they observed by X-ray photoelectron spectroscopy (XPS) the appearance of Mo_2C at 500 °C and the complete transformation of MoO_3 to Mo_2C at 700 °C [30]. Li et al. studied the CHR process using CNT (as-received and nitric-treated) at four different carburization temperatures ranging from 500 °C to 700 °C; Mo_2C formation was observed at a relatively low temperature of 580 °C for both supports. However, the nitric treatment prevented the metallic agglomerations and the destruction of the support after the process even if the CHR was conducted at high temperatures [31]. Although some works have been undertaken using CNF as a catalytic support of Mo_2C nanoparticles [20,23] most of them have been based on carbothermal reduction using N_2, and only a few have focused on CHR. With regard to flowing N_2, Stellwagen and Bitter compared the catalytic activity of CNF-supported W and Mo carbide catalysts in the deoxygenation of stearic acid and related intermediates. An apolar CNF treated with HCl and a polar CNF functionalised with HNO_3 were prepared and used as the supports. The catalyst prepared on the apolar support showed larger crystals compared to the oxidised CNF, which was more stable due to its higher resistance against the metal oxidation [32]. With regard to flowing H_2, Moreira et al. prepared Mo_2C supported on commercial CNF at 750 °C and subsequently used this in the HDO of guaiacol under different operating conditions. They reported the possibility of suffering etching during the CHR when this support is used [33].

The effect of different carbon supports on the formation of Mo_2C during CHR needs to be further studied. Santillana et al. tested three carbon supports (AC, CNT and CNF) with 7.5 Mo wt.% or 20 Mo wt.% in diluted 10% H_2/He at 1000 °C in the HDO of guaiacol. For Mo_2C/AC, only an amorphous Mo phase was observed, in contrast to Mo_2C/MoC crystalline particles for the filamentous carbon. Indeed, due to the well-dispersed Mo_2C nanoparticles on CNF, this catalyst showed the highest selectivity to phenol and catalytic activity at 350 °C [24]. In two different works, Qin et al. compared several carbon supports: AC, CNF, CNT carburised at 700 °C [21]; and AC, graphite, fullerenes and reduced graphene oxide (RGO) carburised at 500–700 °C in CHR [34]. The work optimised the Mo_2C loading in 20 wt.% and compared the three supports in the HDO of methyl palmitate. In this reaction, all the catalysts showed a similar conversion, even though the Mo_2C/CNT showed different product distribution compared to Mo_2C/CNF and Mo_2C/AC. The latter work was focused on the Mo_2C formation in CHR at ranging temperatures of 500–700 °C and their catalytic activity in the HDO of maize oil. Particularly, the catalyst prepared on RGO by means of oxidation and a reduced method showed the highest catalytic activity, enhanced by the active sites and Mo_2C particle size. These authors attributed a catalytic behaviour comparable to noble metal catalysts to the Mo_2C-based catalyst due to C insertion into the Mo lattice, which shows that the d-electron density of the states of the early transition metal becomes higher at the Fermi level [35].

In addition, when Mo_2C is prepared by CHR using carbon materials as a support, the formation of coke on the surface of the catalysts, typically associated with the use of hydrocarbons in the carburization step, is avoided [28]. Normally, Mo_2C supported in metal oxides (TiO_2, ZrO_2, CeO_2, SiO_2 or Al_2O_3) prepared by temperature programmed reduction with hydrocarbons shows deactivation problems due to coke deposition, resulting in a low specific surface area and low stability, and thus low catalytic activity [14].

As described in this brief introduction, CHR is a complex process in which the carbon source is one of the key factors. Depending on the nature of the carbon structure, the CHR conditions may not necessarily remain the same. Therefore, in order to compare the influence of carbon as a Mo_2C support, it is mandatory to determinate the most suitable CHR conditions individually in order to simultaneously preserve the physicochemical properties of the support and achieve well-developed Mo_2C crystal formation. The successful development of carbon-supported Mo_2C catalysts may pave the way for the deployment of HDO technologies, which is of utmost importance to effectively producing high value-added products from biomass resources.

In this work, differently nanostructured carbon nanomaterials (CNF, CNT, RGO and a commercial AC) were used as Mo_2C supports. Within the framework described above for the CHR, it is of utmost importance to carefully select the CHR condition since each carbon support shows its inherent properties in the carburization of Mo, resulting in different β-Mo_2C crystal sizes and affecting considerably the final textural properties of the catalyst. Thus, prior to the carburization process, Mo-impregnated supports were subjected to a thermogravimetric study mimicking the CHR conditions, followed by the characterization by X-ray diffraction of the resulting catalysts. The objective was to determine the optimal CHR temperature, aiming to preserve the support and achieve a well-developed β-Mo_2C crystal structure. Once the carburization temperatures were selected for each support, the catalyst was prepared by CHR in a fixed bed reactor. Subsequently, its catalytic activity was tested in the HDO of guaiacol at relatively mild conditions (300 °C, 20 H_2 bar and 2 h). The guaiacol molecule is widely used as an aromatic model compound of fast pyrolysis oil [36–38]. The catalysts were compared regarding theur guaiacol conversion, product distribution and selectivities, and oxygen removal efficiency.

2. Materials and Methods

2.1. Synthesis of Nanostructured Carbon Materials

Carbon nanofibers were produced in a rotatory bed reactor by $CH_4:CO_2$ (50:50 vol.%) catalytic decomposition at 650 °C using a Ni-Co/Al$_2$O$_3$ catalyst (33.5:33.5:33 molar%). This material was purified with concentrated HCl at 60 °C for 4 h and under mild sonication. After that, the solid was filtered, rinsed with distillate water until a pH of 7 was reached and dried overnight at 60 °C. Subsequently, an oxidation stage was performed with concentrated HNO_3 at boiling point for 1 h and under reflux. After oxidation, the product was filtered, washed and dried as in the previous purification stage. These purified and oxidised carbon nanofibers are referred to as CNF in the manuscript. More details of this synthesis can be found in previous works [18,19].

CNTs were produced by the decomposition of a methane-rich stream (H_2 = 10 L/h; CH_4 = 48 L/h; N_2 = 38 L/h) in a rotatory bed reactor, as elsewhere described [39]. Fe-Mo/MgO (7.5:1.5:91.0 molar %) prepared by co-precipitation was used as a catalyst at 900 °C and for 30 min. The synthesis of these catalysts and the procedures of CNT growth are described elsewhere [40,41]. Additionally, CNTs were purified and oxidised as previously described for CNF.

RGO was synthetised by the chemical unzipping of CNT using a modification of Hummers' method [42] and subsequent hydrothermal reduction. In this case, CNTs were produced in the rotatory bed reactor with a massive Fe-based catalyst, as described in [42]. Aiming to remove the high oxygen content obtained after Hummers' method, the graphene oxide was reduced hydrothermally at 180 °C and for 6 h in an autoclave reactor [43]. After reduction, RGO was dried overnight at 60 °C.

Commercial activated carbon (NORIT C GRAN, CABOT), designed as AC, was used as received. This material was ground in a mortar, and no further treatments were performed. All these carbon materials were grouped under the acronym NCM (nanostructured carbon material).

2.2. Mo Precursor Impregnation and Carbothermal Hydrogen Reduction

Mo was incorporated on the NCM by incipient wetness impregnation using an $(NH_4)_6Mo_7O_{24} \cdot 4H_2O$ aqueous solution (AHM) [18] in order to obtain a Mo content of 10 wt.% after drying (10.6 wt.% of Mo after the transformation of AHM into Mo_2C in CHR process). These materials were labelled as AHM/NCM (where NCM refers to CNF, CNT, RGO or AC). Although the temperature program used in the CHR was established in previous works for CNF-supported catalysts [18,19], the CHR temperature was initially optimised for the different carbon supports by using a thermobalance coupled with a mass spectrometer (TG-MS) (NETZSCH TG 209 F1 Libra® thermobalance coupled to a MICROMERITICS AutoChem™ II 2920 mass spectrometer), where the reduction process was simulated followed by the structural analysis of the resulting catalysts (labelled as Mo_2C/NCM) by X-ray diffraction (XRD). The impregnated NCM was introduced in the thermobalance and carboreduced for 2 h at temperatures ranging from 650 °C to 800 °C, using H_2/Ar (10/90 vol./vol.) as reduction gas and with a temperature ramp of 10 °C/min.

Once the carburization temperature for each support was selected, the CHR of each AHM/NCM was carried out in a fixed-bed tubular quartz reactor using 2.0 g of the sample, 100 mL/min of pure H_2 and atmospheric pressure. The carboreduction program consisted of three steps: fast heating to 350 °C at 10 °C/min, slow heating to the optimal temperature at 1 °C/min [18], and an isothermal period of 1 h at this temperature. Subsequently, the reactor was cooled down under an N_2 atmosphere, and the catalysts were passivated with an O_2/N_2 (1/99 vol./vol.) mixture using a flow rate of 24 mL/min at 25 °C for 2 h. The carboreduced catalysts (reduced to their optimal temperatures) were named as Mo_2C/NCM and used in the HDO of guaiacol.

2.3. Catalytic Hydrodeoxygenation of Guaiacol Using Mo_2C/NCM

The catalytic tests were conducted in an autoclave reactor at 300 °C and 20 bar of H_2 (at room temperature). In a typical run, 1.2 mL of guaiacol was dissolved in 40 mL of n-decane and introduced in the reactor with 0.2 g of catalyst. This mixture was heated to 300 °C at 10 °C/min under soft stirring (300 rpm) to minimize the contact. Once the operating temperature was reached, the reaction was performed for 2 h under vigorous stirring (1000 rpm). The liquid products were filtered and analysed by Gas Chromatography with a Flame Ionization Detector (GC-FID) as in previous works [18].

Briefly, the guaiacol conversion, HDO ratio (parameter which allows comparing the oxygen proportion in the products), specific product yield (SPY) and mass balance were calculated as follows:

$$\text{Conversion (\%)} = \text{(mol of guaiacol in feed} - \text{mol of guaiacol in the product)/mol of guaiacol in feed} \times 100 \qquad (1)$$

$$\text{HDO ratio (\%)} = \text{(mol of O in feed} - \text{mol of O in the product)/ mol of O in feed} \times 100 \qquad (2)$$

$$\text{Mass balance (\%)} = \text{(mass of products} + \text{mass of unreacted guaiacol)/mass of guaiacol in feed} \times 100 \qquad (3)$$

$$\text{SPY (wt.\%)} = \text{(mass of product/mass of Mo in the catalyst)} \times 100 \qquad (4)$$

2.4. Characterization of NCM and Mo₂C/NCM Catalysts

The physical properties of the carbon supports were determined by means of Micrometrics ASAP2020 equipment. Through the 77 K N_2 physisorption, the specific surface area according to the BET method (SBET) and the total pore volume (Vt) at a relative pressure of $P/P_0 > 0.971$ were calculated. Additionally, for RGO, CO_2 physisorption at 273 K was performed in order to determine the micropore surface area by the Dubinin–Radushkevich equation. The residual catalyst of the prepared support and the metallic impurities of AC were determined by thermo-gravimetric analysis by flowing 50 mL/min of air and heating the NCM to 1000 °C at 10 °C/min; the mass loss was recorded in a TG 209 F1 Libra (NETZSCH®). The textural properties of the prepared NCM were observed by transmission electron microscopy (TEM), supporting the NCM in copper grids covered with a lacey amorphous carbon film. TEM images were captured using a JELO-2000 FXII microscope operating at 200 KeV for CNF; for AC, CNT and RGO, a Tecnai F30 (FEI) at 300 KeV was employed. For the identification and size determination of the crystal phases, a DIFRAC PLUS EVA 8.0 diffractometer was used. The identification of the crystal phases was determined by TOPAS software using Rietveld refinement, while the phase identification was realised according to the EVA software package with the International Centre for Diffraction Data database.

The sample scanning was performed at $2\theta = 20°–80°$ by 0.02°/s by means of a Bruker D8 Advance Series 2 diffractometer equipped with Ni-filtered CuKα radiation and a secondary graphite monochromator. The catalyst surface was determined by an ESCAPlus (OMICROM) spectrometer under a $<5 \times 10^{-9}$ Torr vacuum and analysed using CASA XPS software applying a Shirley-type background. The emitted radiation was generated by a hemispherical electron energy analyser, an X-ray at 225 W (15 mA and 15 KV) and a non-monochromatised MgAlα. In order to determine the metallic final composition, the catalyst was treated following the sodium peroxide fusion procedure prior to introduction in a Spectroblue (AMETEK) inductively coupled plasma-optical emission spectrometer (ICP-OES). To obtain the high-angle annular dark field (HAADF) images, the samples were deposited in amorphous carbon-coated copper grids an introduced into a Tecnai F30 (FEI) at 300 KeV using the scanning transmission electron microscope (STEM) mode. The STEM mode allows the use of a coupled energy-dispersive X-ray spectroscope (EDX) for in situ chemical composition analysis.

3. Results and Discussion

3.1. Characterization of NCM

The morphological appearance of the supports was observed by TEM, as shown in Figure 1. CNF showed a graphite stacking in the form of a fishbone-type nanofilament, with an inner hollow core of around 6 nm in diameter and an average outer diameter of 20–30 nm (Figure 1a,b). CNT showed narrower diameters (10–15 nm) with a regular stacking of 8–20 concentric layers (Figure 1c–d). RGO was composed by agglomerates of wrinkled sheets of 4–6 layers (Figure 1e–f). The commercial AC consisted of spherical aggregates composed of randomly aligned small graphite crystallites (Figure 1g–h).

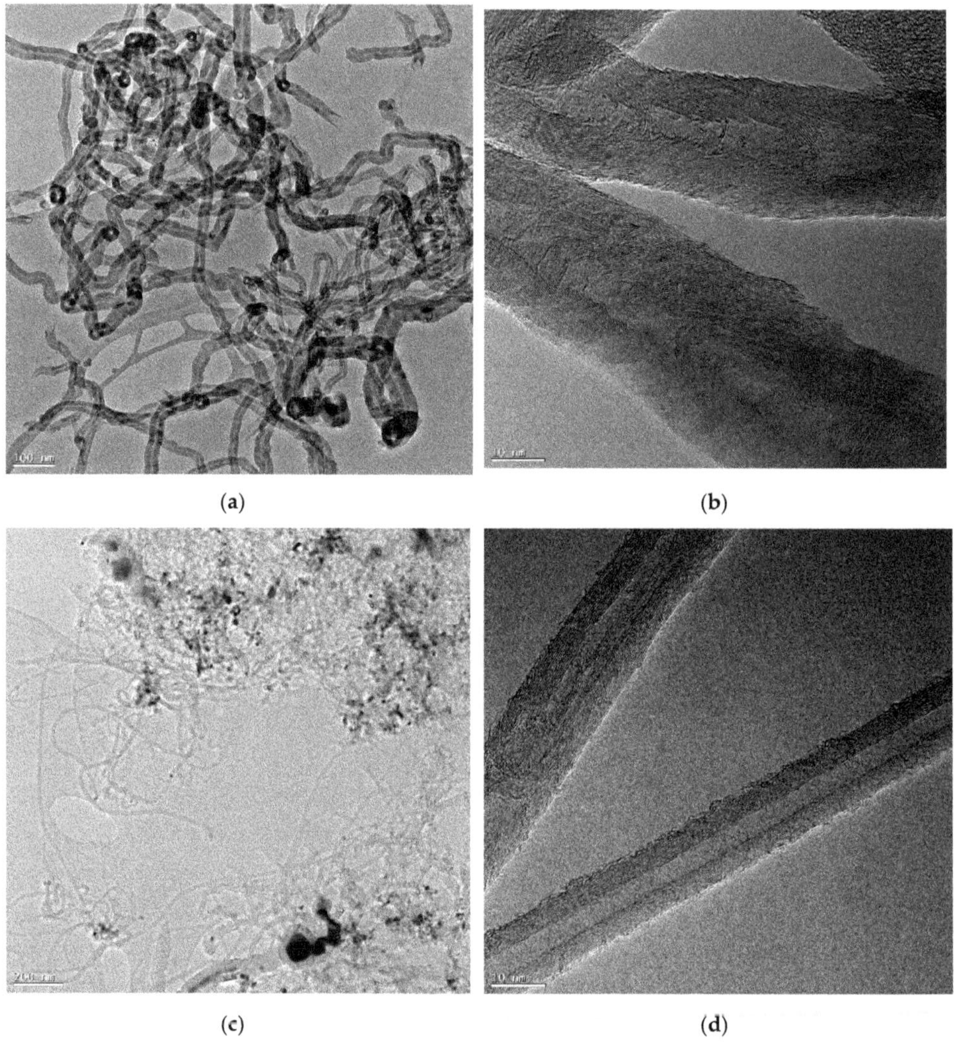

(a)

(b)

(c)

(d)

Figure 1. *Cont.*

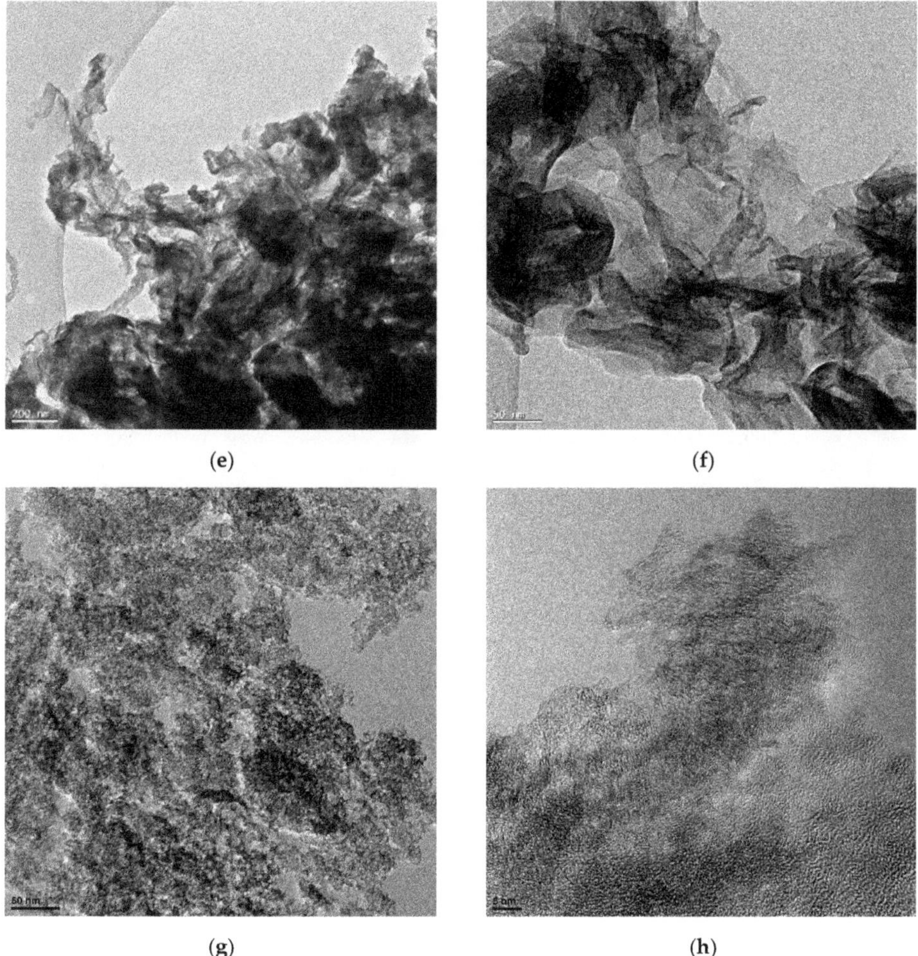

Figure 1. Transmission electron microscope (TEM) images of carbon nanofibers (CNF) (**a,b**), carbon nanotubes (CNT) (**c,d**), reduced graphene oxide (RGO) (**e,f**) and commercial activated carbon (AC) (**g,h**).

In order to determine the surface chemistry of the supports, XPS analysis was performed. The results are summarised in Table 1. All samples are mostly compounds of C and O, related to the functionalities introduced in the oxidation stage, and minor elements (N, P, B, Ni, Co or Al) from their different synthesis routes accounting for below 2 at.%. CNFs could be observed at the surface of Ni, Co and Al, related to the catalysts used in the preparation and which were not completely removed after the oxidation treatment. Commercial AC showed the presence of surface B and P.

The textural properties of carbon supports were determined by N_2 physisorption and are shown in Table 1. AC had a high surface area value (1230 m^2/g), which was higher than carbon nanofilaments (99–104 m^2/g) and RGO (23 m^2/g). Likewise, V_t, ranging from 0.1 cm^3/g to 0.9 cm^3/g, followed the same order as S_{BET}. RGO showed the lowest values of surface area and pore volume, although this material presented a high content of micropores (up to 76 m^2/g of micropore surface area calculated with Dubinin–Radushkevich using CO_2 as adsorbate).

Table 1. Physicochemical properties of nanostructured carbon materials. NCM: nanostructured carbon material; XPS: X-ray photoelectron spectroscopy.

NCM	XPS (at.%)					N$_2$ Physisorption	
	C	O	S	N	Others	S_{BET} (m^2/g)	V_t (cm^3/g)
CNF	93.2	4.8	0.0	0.0	2.0	99	0.463
CNT	95.0	4.3	0.0	0.7	0.0	104	0.590
RGO	70.3	20.5	6.7	2.5	0.0	23	0.105
AC	90.7	7.4	0.0	0.0	1.9	1230	0.913

3.2. Carburization Study in Thermobalance

In order to determine the optimal CHR temperature, the carburization of AHM/NCM samples was carried out in a thermobalance coupled with MS (TG-MS). As previously reported [18], the evolution of m/z = 28 during the CHR stage allowed us to select the temperature necessary to carry out the carburization process. This m/z value is related to the CO formed during the carburization of MoO_2 to Mo_2C. Figure 2 shows this evolution with temperature. The appearance of a CO signal started at different temperatures depending on the support used in the AHM carboreduction. Thus, an m/z = 28 signal appeared at around 500 °C for AHM/AC, at around 560 °C for carbon nanofilaments (CNF and CNT) and at around 630 °C for RGO. An m/z = 28 signal reached a maximum at temperatures centred around 625–645 °C, except for RGO, for which the MoO_2 carburization (and CO formation) was hindered and shifted up to around 725 °C. These differences may be tentatively attributed to the physicochemical properties. Thus, the lower oxygen content in the support resulted in lower CO onset temperature. This factor may influence the dispersion and interaction between the Mo precursor and the support, thus affecting the CO evolution during carbide formation.

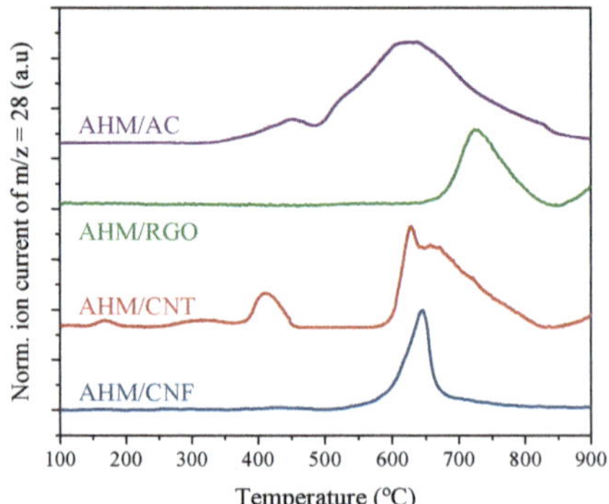

Figure 2. m/z = 28 evolution of the different $(NH_4)_6Mo_7O_{24}\cdot4H_2O$ aqueous solution (AHM)/NCM measured by a thermobalance coupled with mass spectrometer (TG-MS).

The transformation of AHM/NCM (10.0 Mo wt.%) to Mo_2C/NCM entails a theoretical mass loss of 5.4% (AHM→ MoO_3→MoO_2→Mo_2C); nevertheless, all materials exhibited larger mass losses, which were attributed to the gasification of the carbon support (Figure 3a). Gasification is a key parameter to be considered in CHR, since it may modify not only the catalyst porosity but also the final Mo content [18]. Besides this, the CH_4 released during the support gasification may also increase the carburization degree and the β-Mo_2C crystal size. In this way, the weight changes suffered by AHM/NCM materials after an isothermal period of 2 h at different CHR temperatures were analysed (Figure 3a). AHM/CNF suffered mild gasification (13.0–14.3 wt.%) regardless of the final carburization temperature, indicating the larger thermal stability of this material. On the contrary, the gasification in the other NCM was more severe, with the weight loss observed being higher when the carburization temperature was increased. The largest weight loss was observed for the AHM/CNT, ranging from 51.6% to 83.5% when the temperature was increased from 650 °C to 750 °C. AHM/RGO and AHM/AC showed similar behaviours, with weight loss values ranging from 34.4% to 66.1% as the temperature was raised. In the case of RGO, the principal cause may be derived from the removal of the large amount of oxygen functional groups and the subsequent defect generation. Likewise, the thinness of the RGO may favour the fast gasification of this material, contributing to the mass loss. Equally, AC presents a high surface area and a pseudo-amorphous structure, which makes it easier for gasification to occur during CHR.

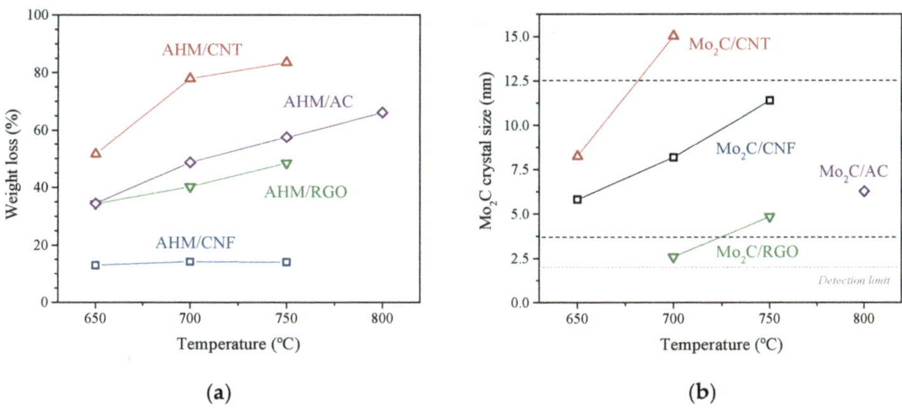

Figure 3. Weight loss (**a**) and β-Mo$_2$C crystal size (**b**) of AHM/NCM and catalysts, respectively, after 2 h carburization in TGA at different temperatures. Dashed lines delimit the optimal β-Mo$_2$C crystal sizes.

After carburization, Mo$_2$C/NCM catalysts were analysed by XRD in order to quantify the size of the β-Mo$_2$C crystal phase (Figure 3b). For temperatures that resulted in crystal sizes below the XRD detection limit, no data are shown. As previous work revealed [18], the optimal β-Mo$_2$C crystal size for the HDO of guaiacol is around 10 nm. Generally, an increase in the carburization temperature led to larger β-Mo$_2$C crystals, although the extension of this growth was markedly different depending on the support. CNT showed larger β-Mo$_2$C crystals than those obtained for RGO and AC at the same temperature: for instance, at 700 °C, the β-Mo$_2$C crystal size of these last catalysts was 2.6 nm and below the detection limit, respectively, in contrast to the 15.0 nm obtained for CNT. The formation of β-Mo$_2$C over CNF showed an intermediate behaviour: 5.8 and 11.4 nm crystals were achieved at 650 and 750 °C, respectively. Likewise, as important as the formation of well-developed β-Mo$_2$C crystals is the mitigation of the gasification of the support. The deep gasification of the support hampers the control of the Mo loading, reducing the surface area of the catalyst and worsening the Mo dispersion on the support surface. Therefore, carbon gasification should be avoided as far as possible to achieve the best catalytic behaviour. It is important to observe that not all NMC tested led to the formation of β-Mo$_2$C crystals of ca. 10 nm. Instead of this, an optimal carburization temperature for each NCM was selected, minimizing the support gasification and aiming to form β-Mo$_2$C crystals as close to 10 nm as possible. In the case of the CNT-supported catalyst, 10 nm β-Mo$_2$C crystals were formed when performing the CHR at 650 °C. Besides this, the selection of this temperature minimizes the gasification of the support (51.6% weight loss). The CHR temperature selected for the CNF-supported catalyst was 750 °C. This support withstood this temperature without suffering a significant mass loss (14.2%) and led to 11.4 nm β-Mo$_2$C crystals. Regarding the optimal CHR temperature for RGO, 750 °C was the temperature selected despite obtaining relatively small β-Mo$_2$C crystals (4.9 nm) and high weight loss (48.4%). Finally, in the case of AC, only temperatures above 800 °C resulted in β-Mo$_2$C crystals over the detection limit, so this temperature was selected as the optimal one, resulting in ca. 6.3 nm crystals and 66.1% weight loss.

3.3. Characterization of Catalysts Using Different NCM

The carburization of AHM/NCM samples was conducted in a fixed-bed reactor at the optimal temperature selected in the previous section. In this case, 100% of H_2 flow and a 1 h isothermal period were employed. Figure 4 and Table 2 show the XRD patterns and β-Mo_2C crystal sizes of the different catalysts. Catalysts showed the typical reflections associated with graphite and the hexagonal close packing structure of β-Mo_2C. AC also showed the presence of MoO_2 and an oxycarbide specie related to MoC_2O_4, which indicates the incomplete carburization of this catalyst. Additional reflections were assigned to support impurities (tge diffractograms of supports are shown in Figure S1 in the Supplementary Information): Mo_2C/CNT contained metallic Fe from the CNT catalyst, and Mo_2C/AC, B and BPO_4 from the AC functionalization treatment. Regarding the β-Mo_2C crystal sizes obtained, they differed from those obtained in the thermobalance (previous section), but this discrepancy was attributed to the H_2 concentration used (10 vol.% vs. 100 vol.%). Pure H_2 flow may facilitate the initial reduction of AHM to MoO_2 [44], hence promoting the carburization of MoO_2 to Mo_2C at lower temperature [28]. Mo_2C is known to act as a gasification catalyst with the concomitant production of CH_4 [9]. This would explain the relatively large β-Mo_2C crystal sizes obtained in the fixed-bed reactor, except for Mo_2C/CNF, which presented a similar β-Mo_2C crystal size than that obtained in TGA. It seems obvious that gasification promotes the sintering of the Mo_2C crystals upon reaction with the CH_4 formed, at least for the AC and CNT-supported catalysts. The gasification of the support during the CHR was further evidenced by the ICP-OES results (Table 2), where a drastic increase of the Mo content was observed for these two catalysts. Taking into consideration the fact that the theoretical Mo value should be around 10.6 wt.% in the final catalyst, the increase in the weight loss observed is entirely provoked by the partial gasification of the support. The fact that the RGO-supported catalyst did not follow this trend could be attributed to the fact that most of the weight loss observed during the CHR stage might be related to the loss of oxygen surface groups rather than carbon gasification.

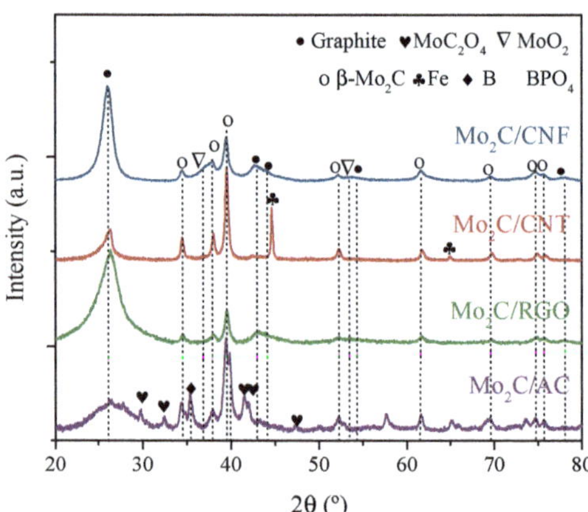

Figure 4. XRD patterns of Mo_2C/NCM catalysts obtained after carbothermal hydrogen reduction (CHR) at the optimum temperature using a heating rate of 1 °C/min and a 1 h isothermal period.

Energies **2020**, 13, 1189

Table 2. Results of characterization of Mo_2C/NCM catalysts obtained in fixed-bed reactor. XRD: X-ray diffraction; ICP-OES: inductively coupled plasma-optical emission spectrometry.

		CHR	XRD (nm)	ICP-OES (wt.%)	XPS (at.%)								N$_2$ Phys.	
Catalyst	T (°C)	Wt. loss (%)	β-Mo$_2$C	Mo	C	O	Mo	Others	Mo^{6+}	Mo^{4+}	Mo$^{\delta+}$	Mo^{2+}	S$_{BET}$ (m^2/g)	V$_t$ (cm^3/g)
Mo$_2$C/CNF	750	30	10.9	12.9	92.6	5.1	2.3	0.01	68.8	2.6	11.7	16.8	64	0.398
Mo$_2$C/CNT	650	70	15.5	25.6	89.7	7.0	3.2	0.14	69.8	3.3	10.4	16.5	106	0.551
Mo$_2$C/RGO	750	58	13.6	13.7	69.2	12.5	6.6	11.8	80.6	1.9	3.1	14.5	40	0.097
Mo$_2$C/AC	800	70	12.5	23.6	85.8	7.9	3.1	3.13	89.6	7.9	0.04	2.5	572	0.653

The surface atomic composition of all catalysts was measured by XPS. XPS survey spectra showed different Mo contents depending on the catalyst (Table 2). The catalysts presented Mo surface compositions in the range of 2.3–3.2 at.%, except for Mo_2C/RGO, which exhibited a higher Mo content (6.6 at.%). In the case of Mo_2C/RGO, 11.8 at.% of S was identified, which was related to the H_2SO_4 used in the RGO synthesis by Hummers' method; no other element, with the exceptions of Mo, C and O, was detected. Finally, the Mo_2C/AC catalyst showed 3.13 at.% of impurities (B and P). On the other hand, the impurities detected at the surface for Mo_2C/CNT and Mo_2C/CNF were negligible. Regarding the oxygen content, a part of the oxygen measured belonged to the O_2 passivation and the formation of MoO_xC_y species [11,34], except for the RGO catalyst, in which the oxygen content is inherent to its structure. Before passivation, CHR removed most of the oxygen groups created in the functionalization of carbon nanofilaments [45].

Aiming to determine the oxidation state of Mo on the catalysts, the deconvolution of the XPS Mo 3d region was undertaken (see Figure S2 in the Supplementary Information). The most oxidised species, Mo^{6+} and Mo^{4+}, are related to MoO_3 and MoO_2, respectively, and are mostly due to the passivation treatment. Several authors reported that the Mo surface is extremely sensitive to oxidation when the sample is in contact with air or O_2 in the passivation treatment [23,32,34,46]. An intermediate oxidation state found between MoO_2 and Mo_2C, named $Mo^{\delta+}$ ($2 < \delta+ < 3$), was associated with oxycarbides [18,19]. Mo^{2+}, located at 228.2–228.4 eV, was related to the Mo_2C specie. Even though the carbide was the desired phase in the catalyst, the results suggested that both species, Mo^{2+} and $Mo^{\delta+}$, may contribute to its hydrodeoxygenation activity [11,18,47]. In any case, the Mo active phase content (Mo^{2+} and $Mo^{\delta+}$) followed this order: $Mo_2C/CNF > Mo_2C/CNT > Mo_2C/RGO >> Mo_2C/AC$.

S_{BET} and V_t of the Mo_2C/NCM catalysts were measured by N_2 physisorption, and the results are listed in Table 2. As a general rule, the catalysts showed smaller S_{BET} and V_t values than their respective supports, motivated by the partial covering of the pores by the Mo phase (Table 1), except for the Mo_2C/GO catalyst which had a slight increase in the porosity values due to the formation of meso and micropores upon the removal of the oxygen and sulphur surface groups, as evidenced in Figure S3 in the Supplementary Information. The AC-based catalyst had a much more pronounced S_{BET} and V_t reduction, attributed to a partial destruction of its micro and mesoporosity (see Figure S3 in the Supplementary Information).

The morphology of all catalysts was observed by STEM. Likewise, the composition of bright particles was determined by EDX (see Figures S4–S6 in the Supplementary Information). Particle size was calculated by measuring the diameter of isolated metallic bright particles from TEM images. In order to do so, more than 250 particles were considered in the analysis. Mo_2C/CNF (Figure 5) showed a β-Mo_2C particle size distribution around 4.9 ± 2.1 nm along with the nanofiber structure. The particle sizes of these nanoparticles shown by the HAADF images are different from those calculated by XRD (10.9 nm). On the other hand, a homogeneous dispersion of nanoparticles lower than 2 nm was observed covering the CNF (Figure 5b). However, these nanoparticles were under the detection limit of EDX and XRD techniques due to their small size. In the case of Mo_2C/CNT (Figure 6), the STEM micrographs revealed some differences in the morphology and composition of particles in the sample. Figure 6a shows a representative image of this material, showing some areas of metal accumulation probably motivated by a heterogeneous impregnation of the precursor (white rectangle). These areas coexisted with the presence of long CNTs decorated with β-Mo$_2$C particles of different sizes both in the inner and in the outer surface (white circles) and iron particles from the catalyst employed on the support growth embedded in the inner core (white arrows), as determined by EDX (see Figure S4 in the Supplementary Information). In the case of Mo$_2$C/RGO (Figure 7), this catalyst showed a high concentration of well-dispersed small metallic nanoparticles (2–4 nm). Besides, the chemical composition of bright particles was related exclusively to Mo species as shown in Figure S5 (see Supplementary Information). Similarly, Mo$_2$C/AC showed a high metal concentration covering the support (Figure 8). According to the performed EDX (see Figure S6 in the Supplementary Information), the support was comprised of Mo, C and P, with the bright round nanoparticles being the dispersed Mo$_2$C.

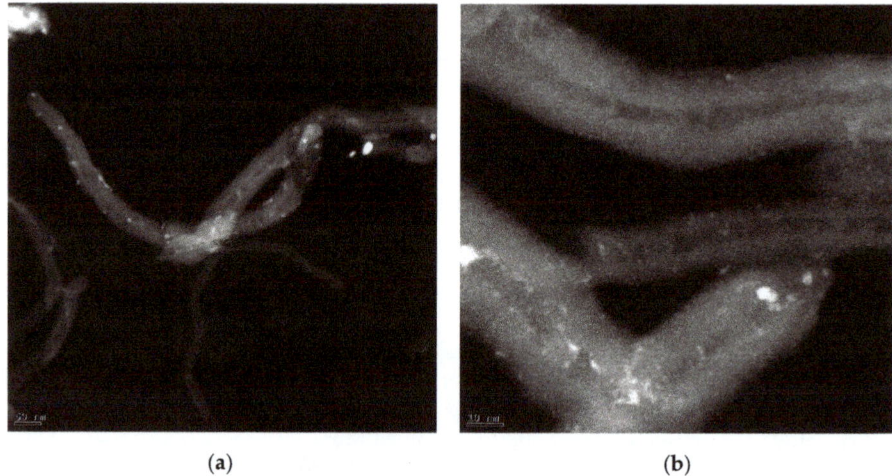

(a)   (b)

Figure 5. Low (**a**) and high-magnification (**b**) STEM images of Mo$_2$C/CNF.

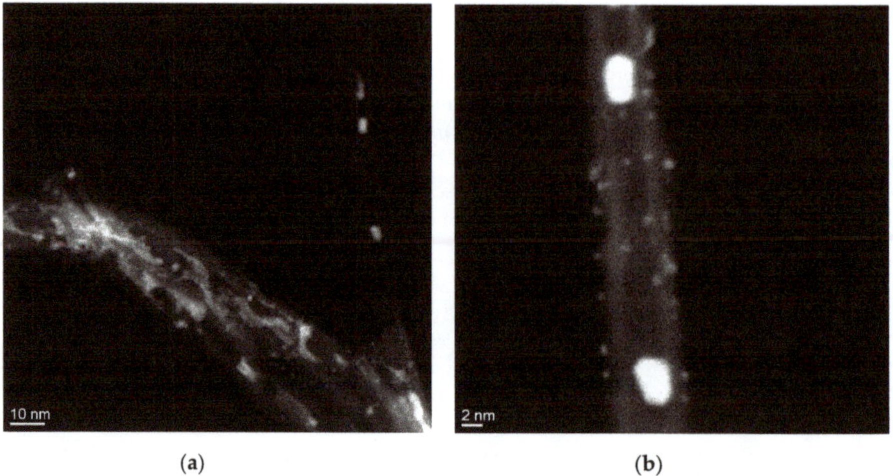

(a)   (b)

Figure 6. Low (**a**) and high-magnification (**b**) STEM images of Mo$_2$C/CNT.

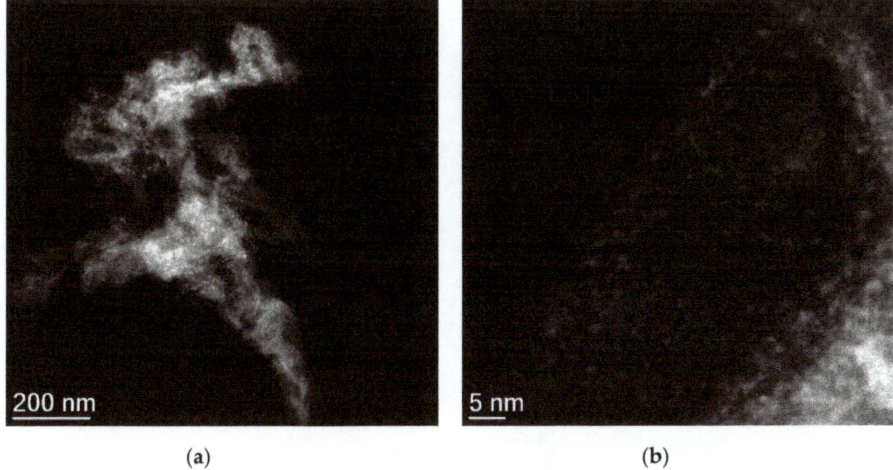

(a)   (b)

Figure 7. Low (**a**) and high-magnification (**b**) STEM images of Mo_2C/RGO.

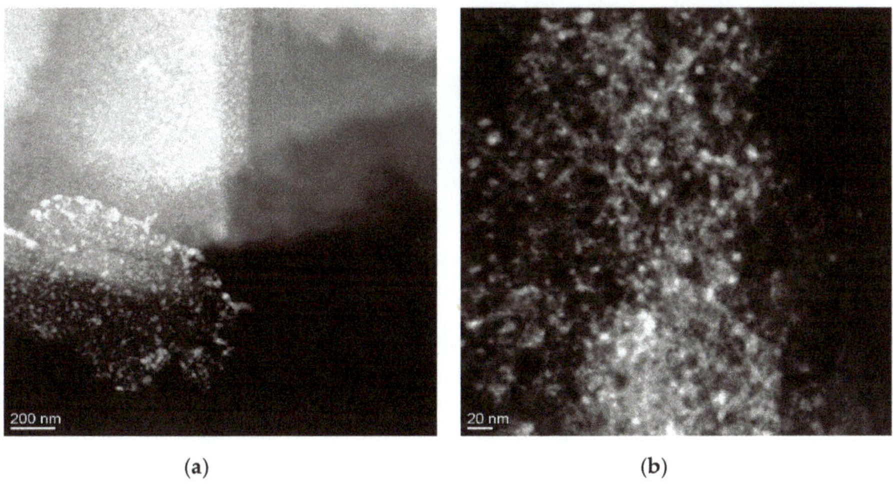

(a)   (b)

Figure 8. Low (**a**) and high-magnification (**b**) STEM images of Mo_2C/AC.

3.4. Catalytic HDO of Guaiacol

Catalysts prepared by CHR were tested in the HDO of guaiacol: Mo_2C/CNF carboreduced at 750 °C, Mo_2C/CNT at 650 °C, Mo_2C/RGO at 750 °C and Mo_2C/AC at 800 °C. As previously stated, the temperatures used in the carburization of each catalyst correspond to the optimum temperature for each support as determined in the previous section. The reaction was performed at 300 °C, 20 H_2 bar measured at room temperature and 2 h of reaction time. The results for each catalyst are summarised in Table 3 and Figure 9. Overall, the catalytic activity (guaiacol conversion and HDO ratio) followed this order: $Mo_2C/CNF > Mo_2C/AC > Mo_2C/RGO > Mo_2C/CNT$. With regard to the characterization results, the β-Mo_2C crystal size measured by XRD, which for this set of catalysts ranged from 10.9 to 15.5 nm, was inversely proportional to the guaiacol conversion. However, this feature cannot solely explain the catalytic activity. A plethora of other parameters, such as the textural properties and the Mo content in the bulk and at the surface, as well as the Mo and composition, have to be also taken into consideration. For instance, the Mo content determined by ICP was 12.9–13.7 wt.% for the CNF and RGO-supported catalysts, where it almost doubled in the case of AC and CNT-based catalysts as a result of the different gasification extents.

Table 3. The conversion, hydrodeoxygenation (HDO) ratio and mass balance obtained for all Mo_2C/NCM catalysts in the HDO of guaiacol at 300 °C, 20 H_2 bar and under a 2 h reaction.

Catalyst	Conversion (%)	HDO Ratio (%)	Mass Balance (%)
Mo_2C/CNF	67.0	37.20	83.28
Mo_2C/CNT	14.23	9.09	94.34
Mo_2C/RGO	20.68	9.59	99.99
Mo_2C/AC	42.00	24.97	83.72

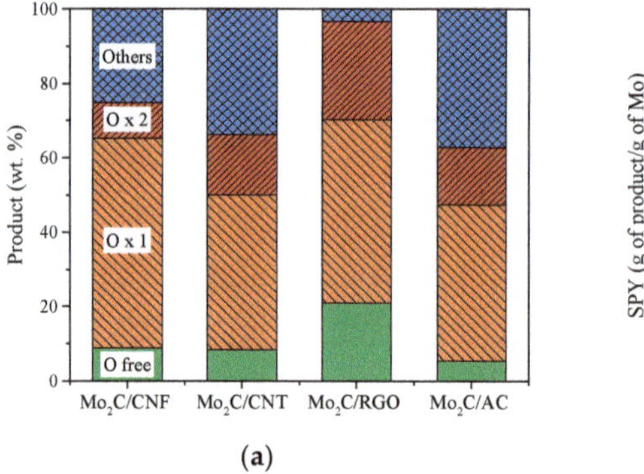

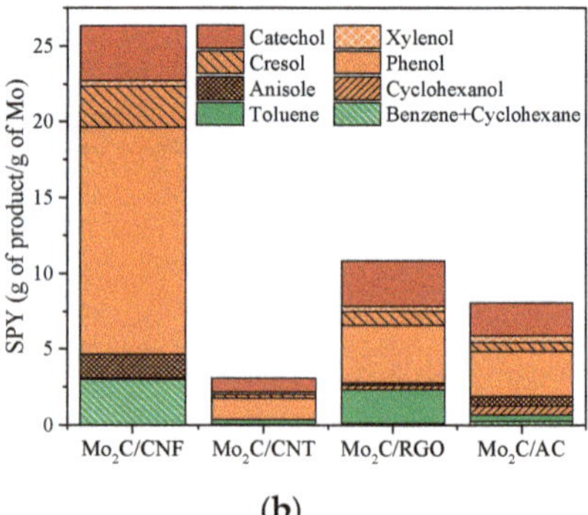

Figure 9. Catalytic results obtained in the HDO of guaiacol using different NCMs as a support of Mo_2C: (**a**) product distribution grouped by oxygen content; (**b**) product production per gram of Mo (SPY).

Mo_2C/CNF showed the best catalytic performance towards the HDO of guaiacol: 67.0% conversion and a HDO ratio of 37.2%. Although Ni and Co traces were detected for this support, the catalytic contribution of these metals were low due to their small proportion and localization [18]. Thus, the catalytic activity of this catalyst can be entirely attributed to Mo^{2+} and Mo^{δ} species, which according to the literature are the most active Mo phases [11]. Mo_2C/CNF showed the largest values, suffering from very soft gasification, and hence the morphology of the support was mostly preserved. Mo_2C/AC followed in terms of the conversion and HDO ratio (42.0% and 25.0%, respectively). It should be noted that this catalyst exhibited a high Mo concentration (24 wt.%) with a similar β-Mo_2C crystal size to Mo_2C/CNF (12.4 nm). The HAADF-STEM images showed a high concentration of metallic particles without agglomerations covering the AC surface. Nonetheless, the low values of surface Mo active phases determined by XPS are striking. Beside this, the microporous nature of this material (see Figure S3 in the Supplementary Information) may account for the lower catalytic activity as compared to the CNF-supported catalysts, despite the larger amount of Mo. The active sites in CNF are much more accessible since they are composed mainly of meso and macropores. A plausible explanation for this observation is that part of the β-Mo_2C in AC is deposited in the micropores, and its activity is hindered by the diffusion constraints of the relatively bulky guaiacol molecule.

Mo_2C/RGO showed low catalytic activity (20.7% conversion and a 9.6% HDO ratio). According to the TEM study, it is clear that this material is mostly composed of very finely dispersed small Mo particles whose oxidation state is not possible to elucidate by this technique. XRD showed the presence of β-Mo_2C, although the small size of the diffraction peak showed that the contribution of this phase is relatively low. Therefore, this catalyst seems to be formed mostly by Mo oxide nanoparticles with a low catalytic activity towards HDO, hence explaining the low activity observed.

Mo_2C/CNT showed the poorest catalytic activity and HDO ratio (14.2% and 9.1%, respectively), even though the catalyst exhibited relatively large Mo_2C crystals (15.5 nm), a similar concentration of Mo active phases in the surface as compared to the Mo_2C/CNF catalyst and a relatively high S_{BET} (105.9 m^2/g). This phenomenon may be tentatively explained by the partial destruction of the support morphology, as evidenced the TEM study, highlighting the importance of the support in not only dispersing the active phase but also possibly serving as a hydrogen reservoir for the hydrogenation reactions [48]. In addition, the exposure of the iron particles that were originally embedded on the inner part of the tubes upon the excessive gasification that takes place during the CHR may also have affected the catalytic behaviour due to the competitive adsorption of guaiacol molecules in Fe and Mo particles.

In addition to the catalytic activity, the product distribution may also be affected by the support used in the catalyst synthesis, as is shown in Figure 9a. The product distribution was split into three categories according to the number of oxygen atoms: 2 (catechol), 1 (phenol, anisole, cresol, xylenol and methyl-cyclohexanol) and 0 (toluene and benzene/cyclohexane. All catalysts showed a higher production of one-oxygen products ($O \times 1$) as a consequence of the removal of one oxygen atom. Likewise, phenol was the main product obtained by the remotion of $-OCH_3$ (see Table S1 in the Supplementary Information).

All catalysts allowed guaiacol to transform into O-free products (under the reaction scheme suggested in previous works [18,48]), and the content was increased for the catalysts with larger Mo_2C crystals. However, only Mo_2C/CNF and Mo_2C/CNT showed a relevant concentration of these compounds in the final liquid (ca. 10%). Despite this, both show a similar catalytic distribution; in the case of O-free products, the Mo_2C/CNF favoured the cyclohexane +benzene production in contrast to Mo_2C/CNT, which exhibited toluene as the main O-free product. Likewise, Mo_2C/RGO showed the highest 0-Oxy concentration due to its selectivity to toluene. Toluene may be formed by the dihydroxylation of cresol or as a consequence of partial deoxygenation from anisole. Regarding the product distribution, the non-detection of anisole for Mo_2C/CNT may be due to the fast conversion of toluene, although a relative proportion of it is identified in Mo_2C/AC and Mo_2C/RGO (2.25% and 0.34%, respectively).

On the other hand, cyclohexane + benzene could be formed by the demethoxylation of anisole or dihydroxylation of phenol; both reaction paths may take place, because both compounds were detected. Particularly, the Mo_2C/CNF catalyst showed a relatively high concentration of those compounds in the final liquid (10%); this fact probably was due to the good catalytic activity shown (with a guaiacol conversion of 66.98%) in the HDO of guaiacol, which led to the reaction achieving the largest product formation.

Catechol was the only product detected by GC which did not show any oxygen removal (2-O product). This product is considered to be an intermediate in the phenol production, and it may form via the cleavage of the $O-CH_3$ bond from guaiacol [47]. Some authors conclude that the formation of 2-O was favoured by the support acidity [49]. In contrast to this hypothesis, some authors attributed higher 2-Oxy production to acidic/basic nature of Mo_2C, which is created by the Mo-C mass transfer [50]. For both cases, the support plays a crucial role in the catalytic behaviour of the active Mo_2C sites. Regarding the 2-O product concentration, higher production was obtained for RGO, which had the largest value of oxygen content and hence the largest acidity, followed by Mo_2C/CNT and Mo_2C/AC. Nonetheless, the possibility that catalytic activity may be affected by other factors such as Mo content, β-Mo_2C crystal size or $Mo^{2+}/Mo^{\delta+}$ ratio cannot be discarded.

The selectivity to the most important and relevant deoxygenated products is shown in Table S2, where phenol, cyclohexane + benzene and toluene were listed. All catalysts were more selective to phenol (1-Oxy product), although there were differences in 0-Oxy selectivity. The CNF-supported catalyst showed higher cyclohexane + benzene selectivity as compared to CNT and RGO-supported catalysts, which were more selective to toluene.

The last group of products plotted was named "Others"; these compounds were calculated by the mass difference between the conversion and the final mass obtained in the identified products. These high molecular compounds were solubilised in the liquid phase but could not be detected by

GC [20]. The production of "Others" was augmented as long as conversion increased (see Table S1 in the Supplementary Information), although their final content was masked by the increase of the formation of detected products with the guaiacol conversion. For that reason, the Mo_2C/AC showed a higher "Others" concentration than Mo_2C/CNF when the production in both reaction tests was similar.

Finally, catalysts were compared according to the specific product yield (SPY), as previously reported in [19] (Figure 9b), which considered the amount of Mo present in the catalyst. Mo_2C/CNF clearly exhibited the highest total yield (as the sum of SPY) compared to the rest of the catalysts. In turn, Mo_2C/AC showed a lower total yield than Mo_2C/RGO, which means that the catalytic activity of the former is more related to the high amount of Mo than to the effectiveness of the active phase. Mo_2C/CNT shows the lowest SPY as a consequence of the possible Fe competition with Mo_2C.

Considering all factors, Mo_2C/CNF showed the best catalytic activity, with good selectivity to oxygen-free HDO products and high guaiacol conversion. This performance was related to its higher gasification resistance, good Mo_2C dispersion and crystal formation in CHR, which makes it the most suitable CNM support for Mo_2C-based catalysts.

4. Conclusions

The nanostructured carbon materials had a direct effect on the formation of β-Mo_2C in the carbothermal hydrogen reduction and hence in the hydrodeoxygenation of guaiacol. The carbon nanofiber-based catalyst resulted in the best catalytic performance, obtaining the highest conversion and yield of desired products (mainly phenol, benzene and cyclohexane). This behaviour was attributed to the higher thermal stability of CNF, which prevented its gasification and promoted a good evolution of the crystal size of Mo species. This catalyst exhibited well-dispersed β-Mo_2C nanoparticles of ca. 11 nm. On the contrary, the other supports suffered from severe gasification (60–70% wt. loss), which resulted in poorer HDO efficiency catalysts regardless of the β-Mo_2C crystal size. This demonstrated the importance of the stability of carbon supports in the Mo_2C-based catalyst prepared by CHR.

Supplementary Materials: The following are available online at http://www.mdpi.com/1996-1073/13/5/1189/s1. Figure S1: XRD patterns of supports after their corresponding purification and functionalization treatments; Figure S2: Mo 3d deconvolution of the Mo_2C/NMC catalysts; Figure S3: DFT Pore size distributions of catalysts and supports measured by N_2 physisorption; Figure S4: EDX performed to Mo_2C/CNT; Figure S5: EDX performed to Mo_2C/RGO; Figure S6: EDX performed to Mo_2C/AC. Table S1: Product distribution (wt.%). Table S2: Selectivity (mol %).

Author Contributions: Conceptualization, J.L.P. and I.S.; methodology, E.O.; validation, D.T. and J.L.P.; formal analysis, E.O.; data curation: E.O.; investigation, E.O.; Supervision, J.L.P. and I.S.; writing—Original draft preparation, E.O.; writing—Review and editing, D.T., J.L.P. and I.S.; funding acquisition, J.L.P. and I.S. All authors have read and agreed to the published version of the manuscript.

Funding: This work was funded by European Regional Development Fund and the Spanish Economy and Competitiveness Ministry (MINECO) (ENE2017-83854-R).

Acknowledgments: EO is grateful for the award of her PhD under the frame of the aforementioned project. JLP thanks CSIC for the financial support (PII 201850I039 project). The authors gratefully acknowledge the "Laboratorio de Microscopías Avanzadas" at "Instituto de Nanociencia de Aragón - Universidad de Zaragoza" for offering access to their microscope and expertise.

Conflicts of Interest: The authors declare no conflict of interest. The funders had no role in the design of the study; in the collection, analyses, or interpretation of data; in the writing of the manuscript, or in the decision to publish the results.

References

1. Volpe, L.; Boudart, M. Compounds of molybdenum and tungsten with high specific surface area: II. Carbides. *J. Solid State Chem.* **1985**, *59*, 348–356. [CrossRef]
2. Dhandapani, B.; Clair, T.S.; Oyama, S.T. Simultaneous hydrodesulfurization, hydrodeoxygenation, and hydrogenation with molybdenum carbide. *Appl. Catal. A: Gen.* **1998**, *168*, 219–228. [CrossRef]

3. Claridge, J.B.; York, A.P.E.; Brungs, A.J.; Green, M.L.H. Study of the Temperature-Programmed Reaction Synthesis of Early Transition Metal Carbide and Nitride Catalyst Materials from Oxide Precursors. *Chem. Mater.* **2000**, *12*, 132–142. [CrossRef]
4. Xiao, T.-C.; York, A.P.E.; Williams, V.C.; Al-Megren, H.; Hanif, A.; Zhou, X.-Y.; Green, M.L.H. Preparation of Molybdenum Carbides Using Butane and Their Catalytic Performance. *Chem. Mater.* **2000**, *12*, 3896–3905. [CrossRef]
5. Brungs, A.J.; York, A.P.E.; Claridge, J.B.; Márquez-Alvarez, C.; Green, M.L.H. Dry reforming of methane to synthesis gas over supported molybdenum carbide catalysts. *Catal. Lett.* **2000**, *70*, 117–122. [CrossRef]
6. Xiao, T.; Hanif, A.; York, A.P.E.; Sloan, J.; Green, M.L.H. Study on preparation of high surface area tungsten carbides and phase transition during the carburisation. *Phys. Chem. Chem. Phys.* **2002**, *4*, 3522–3529. [CrossRef]
7. Hanif, A.; Xiao, T.; York, A.P.E.; Sloan, J.; Green, M.L.H. Study on the Structure and Formation Mechanism of Molybdenum Carbides. *Chem. Mater.* **2002**, *14*, 1009–1015. [CrossRef]
8. Ramanathan, S.; Oyama, S.T. New Catalysts for Hydroprocessing: Transition Metal Carbides and Nitrides. *J. Phys. Chem.* **1995**, *99*, 16365–16372. [CrossRef]
9. Mordenti, D.; Brodzki, D.; Djéga-Mariadassou, G. New Synthesis of Mo_2C 14 nm in Average Size Supported on a High Specific Surface Area Carbon Material. *J. Solid State Chem.* **1998**, *141*, 114–120. [CrossRef]
10. Liang, C.; Ying, P.; Li, C. Nanostructured β-Mo_2C Prepared by Carbothermal Hydrogen Reduction on Ultrahigh Surface Area Carbon Material. *Chem. Mater.* **2002**, *14*, 3148–3151. [CrossRef]
11. Wang, H.; Liu, S.; Smith, K.J. Synthesis and Hydrodeoxygenation Activity of Carbon Supported Molybdenum Carbide and Oxycarbide Catalysts. *Energy Fuels* **2016**, *30*, 6039–6049. [CrossRef]
12. Wang, H.; Liu, S.; Liu, B.; Montes, V.; Hill, J.M.; Smith, K.J. Carbon and Mo transformations during the synthesis of mesoporous Mo_2C/carbon catalysts by carbothermal hydrogen reduction. *J. Solid State Chem.* **2018**, *258*, 818–824. [CrossRef]
13. Liang, C.; Wei, Z.; Xin, Q.; Li, C. Synthesis and characterization of nanostructured Mo_2C on carbon material by carbothermal hydrogen reduction. In *Studies in Surface Science and Catalysis*; Gaigneaux, E., de Vos, D.E., Grange, P., Jacobs, P.A., Martens, J.A., Ruiz, P., Poncelet, G., Eds.; Elsevier: Amsterdam, The Netherlands, 2000; pp. 975–982.
14. Wang, H.; Liu, S.; Govindarajan, R.; Smith, K.J. Preparation of Ni-Mo_2C/carbon catalysts and their stability in the HDS of dibenzothiophene. *Appl. Catal. A: Gen.* **2017**, *539*, 114–127. [CrossRef]
15. Souza Macedo, L.; Stellwagen, D.R.; Teixeira da Silva, V.; Bitter, J.H. Stability of Transition-metal Carbides in Liquid Phase Reactions Relevant for Biomass-Based Conversion. *ChemCatChem* **2015**, *7*, 2816–2823. [CrossRef]
16. Lee, J.S.; Oyama, S.T.; Boudart, M. Molybdenum carbide catalysts. *J. Catal.* **1987**, *106*, 125–133. [CrossRef]
17. Guil-López, R.; Nieto, E.; Botas, J.A.; Fierro, J.L.G. On the genesis of molybdenum carbide phases during reduction-carburization reactions. *J. Solid State Chem.* **2012**, *190*, 285–295. [CrossRef]
18. Ochoa, E.; Torres, D.; Moreira, R.; Pinilla, J.L.; Suelves, I. Carbon nanofiber supported Mo_2C catalysts for hydrodeoxygenation of guaiacol: The importance of the carburization process. *Appl. Catal. B* **2018**, *239*, 463–474. [CrossRef]
19. Ochoa, E.; Torres, D.; Pinilla, J.L.; Suelves, I. Influence of carburization time on the activity of Mo_2C/CNF catalysts for the HDO of guaiacol. *Catal. Today* **2019**. [CrossRef]
20. Jongerius, A.L.; Gosselink, R.W.; Dijkstra, J.; Bitter, J.H.; Bruijnincx, P.C.A.; Weckhuysen, B.M. Carbon Nanofiber Supported Transition-Metal Carbide Catalysts for the Hydrodeoxygenation of Guaiacol. *ChemCatChem* **2013**, *5*, 2964–2972. [CrossRef]
21. Qin, Y.; Chen, P.; Duan, J.; Han, J.; Lou, H.; Zheng, X.; Hong, H. Carbon nanofibers supported molybdenum carbide catalysts for hydrodeoxygenation of vegetable oils. *RSC Adv.* **2013**, *3*, 17485–17491. [CrossRef]
22. Han, J.; Duan, J.; Chen, P.; Lou, H.; Zheng, X. Molybdenum Carbide-Catalyzed Conversion of Renewable Oils into Diesel-like Hydrocarbons. *Adv. Synth. Catal.* **2011**, *353*, 2577–2583. [CrossRef]
23. Hollak, S.A.W.; Gosselink, R.W.; van Es, D.S.; Bitter, J.H. Comparison of Tungsten and Molybdenum Carbide Catalysts for the Hydrodeoxygenation of Oleic Acid. *ACS Catal.* **2013**, *3*, 2837–2844. [CrossRef]
24. Santillan-Jimenez, E.; Perdu, M.; Pace, R.; Morgan, T.; Crocker, M. Activated Carbon, Carbon Nanofiber and Carbon Nanotube Supported Molybdenum Carbide Catalysts for the Hydrodeoxygenation of Guaiacol. *Catalysts* **2015**, *5*, 424. [CrossRef]

25. Liu, S.; Wang, H.; Smith, K.J.; Kim, C.S. Hydrodeoxygenation of 2-Methoxyphenol over Ru, Pd, and Mo$_2$C Catalysts Supported on Carbon. *Energy Fuels* **2017**, *31*, 6378–6388. [CrossRef]
26. Pielaszek, J.; Mierzwa, B.; Medjahdi, G.; Marêché, J.F.; Puricelli, S.; Celzard, A.; Furdin, G. Molybdenum carbide catalyst formation from precursors deposited on active carbons: XRD studies. *Appl. Catal. A: Gen.* **2005**, *296*, 232–237. [CrossRef]
27. He, L.; Qin, Y.; Lou, H.; Chen, P. Highly dispersed molybdenum carbide nanoparticles supported on activated carbon as an efficient catalyst for the hydrodeoxygenation of vanillin. *RSC Adv.* **2015**, *5*, 43141–43147. [CrossRef]
28. Frank, B.; Friedel, K.; Girgsdies, F.; Huang, X.; Schlögl, R.; Trunschke, A. CNT-Supported Mo$_x$C Catalysts: Effect of Loading and Carburization Parameters. *ChemCatChem* **2013**, *5*, 2296–2305. [CrossRef]
29. Frank, B.; Xie, Z.-L.; Ortega, K.F.; Scherzer, M.; Schlogl, R.; Trunschke, A. Modification of the carbide microstructure by N- and S-functionalization of the support in Mo$_x$C/CNT catalysts. *Catal. Sci. Technol.* **2016**, *6*, 3468–3475. [CrossRef]
30. Barthos, R.; Széchenyi, A.; Solymosi, F. Efficient H$_2$ Production from Ethanol over Mo$_2$C/C Nanotube Catalyst. *Catal. Lett.* **2008**, *120*, 161–165. [CrossRef]
31. Li, X.; Ma, D.; Chen, L.; Bao, X. Fabrication of molybdenum carbide catalysts over multi-walled carbon nanotubes by carbothermal hydrogen reduction. *Catal. Lett.* **2007**, *116*, 63–69. [CrossRef]
32. Stellwagen, D.R.; Bitter, J.H. Structure-performance relations of molybdenum- and tungsten carbide catalysts for deoxygenation. *Green Chem.* **2015**, *17*, 582–593. [CrossRef]
33. Moreira, R.; Ochoa, E.; Pinilla, J.; Portugal, A.; Suelves, I. Liquid-Phase Hydrodeoxygenation of Guaiacol over Mo$_2$C Supported on Commercial CNF. Effects of Operating Conditions on Conversion and Product Selectivity. *Catalysts* **2018**, *8*, 127. [CrossRef]
34. Qin, Y.; He, L.; Duan, J.; Chen, P.; Lou, H.; Zheng, X.; Hong, H. Carbon-Supported Molybdenum-Based Catalysts for the Hydrodeoxygenation of Maize Oil. *ChemCatChem* **2014**, *6*, 2698–2705. [CrossRef]
35. Choi, J.-S.; Bugli, G.; Djéga-Mariadassou, G. Influence of the Degree of Carburization on the Density of Sites and Hydrogenating Activity of Molybdenum Carbides. *J. Catal.* **2000**, *193*, 238–247. [CrossRef]
36. He, Z.; Wang, X. Hydrodeoxygenation of model compounds and catalytic systems for pyrolysis bio-oils upgrading. *Catal. Sustain. Energy* **2012**, *1*, 28–52. [CrossRef]
37. Roldugina, E.A.; Naranov, E.R.; Maximov, A.L.; Karakhanov, E.A. Hydrodeoxygenation of guaiacol as a model compound of bio-oil in methanol over mesoporous noble metal catalysts. *Appl. Catal. A: Gen.* **2018**, *553*, 24–35. [CrossRef]
38. Zhao, H.Y.; Li, D.; Bui, P.; Oyama, S.T. Hydrodeoxygenation of guaiacol as model compound for pyrolysis oil on transition metal phosphide hydroprocessing catalysts. *Appl. Catal. A* **2011**, *391*, 305–310. [CrossRef]
39. Pinilla, J.L.; Utrilla, R.; Lázaro, M.J.; Suelves, I.; Moliner, R.; Palacios, J.M. A novel rotary reactor configuration for simultaneous production of hydrogen and carbon nanofibers. *Int. J. Hydrog. Energy* **2009**, *34*, 8016–8022. [CrossRef]
40. Torres, D.; Pinilla, J.L.; Suelves, I. Unzipping of multi-wall carbon nanotubes with different diameter distributions: Effect on few-layer graphene oxide obtention. *Appl. Surf. Sci.* **2017**, *424*, 101–110. [CrossRef]
41. Torres, D.; Pinilla, J.L.; Lázaro, M.J.; Moliner, R.; Suelves, I. Hydrogen and multiwall carbon nanotubes production by catalytic decomposition of methane: Thermogravimetric analysis and scaling-up of Fe–Mo catalysts. *Int. J. Hydrog. Energy* **2014**, *39*, 3698–3709. [CrossRef]
42. Torres, D.; Pinilla, J.L.; Moliner, R.; Suelves, I. On the oxidation degree of few-layer graphene oxide sheets obtained from chemically oxidized multiwall carbon nanotubes. *Carbon* **2015**, *81*, 405–417. [CrossRef]
43. Torres, D.; Pinilla, J.L.; Galvez, E.M.; Suelves, I. Graphene quantum dots from fishbone carbon nanofibers. *RSC Adv.* **2016**, *6*, 48504–48514. [CrossRef]
44. Ciembroniewicz, G.; Dziembaj, R.; Kalicki, R. Thermal decomposition of Ammonium paramolybdate (APM). *J. Therm. Anal.* **1983**, *27*, 125–138. [CrossRef]
45. Kundu, S.; Wang, Y.; Xia, W.; Muhler, M. Thermal Stability and Reducibility of Oxygen-Containing Functional Groups on Multiwalled Carbon Nanotube Surfaces: A Quantitative High-Resolution XPS and TPD/TPR Study. *J. Phys. Chem. C* **2008**, *112*, 16869–16878. [CrossRef]
46. Liu, C.; Lin, M.; Fang, K.; Meng, Y.; Sun, Y. Preparation of nanostructured molybdenum carbides for CO hydrogenation. *RSC Adv.* **2014**, *4*, 20948–20954. [CrossRef]

47. Chang, J.; Danuthai, T.; Dewiyanti, S.; Wang, C.; Borgna, A. Hydrodeoxygenation of Guaiacol over Carbon-Supported Metal Catalysts. *ChemCatChem* **2013**, *5*, 3041–3049. [CrossRef]
48. Pinilla, J.L.; García, A.B.; Philippot, K.; Lara, P.; García-Suárez, E.J.; Millan, M. Carbon-supported Pd nanoparticles as catalysts for anthracene hydrogenation. *Fuel* **2014**, *116*, 729–735. [CrossRef]
49. Lu, M.; Lu, F.; Zhu, J.; Li, M.; Zhu, J.; Shan, Y. Hydrodeoxygenation of methyl stearate as a model compound over Mo_2C supported on mesoporous carbon. *React. Kinet. Mech. Catal.* **2015**, *115*, 251–262. [CrossRef]
50. Bej, S.K.; Bennett, C.A.; Thompson, L.T. Acid and base characteristics of molybdenum carbide catalysts. *Appl. Catal. A: Gen.* **2003**, *250*, 197–208. [CrossRef]

Article

Ni Supported on Natural Clays as a Catalyst for the Transformation of Levulinic Acid into γ-Valerolactone without the Addition of Molecular Hydrogen

Adrián García [1], Rut Sanchis [1], Francisco J. Llopis [1], Isabel Vázquez [1], María Pilar Pico [2], María Luisa López [3], Inmaculada Álvarez-Serrano [3] and Benjamín Solsona [1,*]

[1] Departament d'Enginyeria Química, ETSE, Universitat de València, Av. Universitat, Burjassot, 46100 Valencia, Spain; adrian.garcia@uv.es (A.G.); rut.sanchis@uv.es (R.S.); francisco.llopis@uv.es (F.J.L.); Isabel.vazquez@uv.es (I.V.)

[2] Sepiolsa, Avda. del Acero, 14-16, Pol. UP-1 (Miralcampo), 19200 Azuqueca de Henares, Spain; maria.pico@sepiolsa.com

[3] Departamento de Química Inorgánica, Facultad de Ciencias Químicas, Universidad Complutense de Madrid, 28040 Madrid, Spain; marisal@ucm.es (M.L.L.); ias@ucm.es (I.Á.-S.)

* Correspondence: benjamin.solsona@uv.es; Tel.: +34-963543735

Received: 11 May 2020; Accepted: 1 July 2020; Published: 3 July 2020

Abstract: γ-Valerolactone (GVL) is a valuable chemical that can be used as a clean additive for automotive fuels. This compound can be produced from biomass-derived compounds. Levulinic acid (LA) is a compound that can be obtained easily from biomass and it can be transformed into GVL by dehydration and hydrogenation using metallic catalysts. In this work, catalysts of Ni (a non-noble metal) supported on a series of natural and low-cost clay-materials have been tested in the transformation of LA into GVL. Catalysts were prepared by a modified wet impregnation method using oxalic acid trying to facilitate a suitable metal dispersion. The supports employed are attapulgite and two sepiolites with different surface areas. Reaction tests have been undertaken using an aqueous medium at moderate reaction temperatures of 120 and 180 °C. Three types of experiments were undertaken: (i) without H_2 source, (ii) using formic acid (FA) as hydrogen source and (iii) using Zn in order to transform water in hydrogen through the reaction $Zn + H_2O \rightarrow ZnO + H_2$. The best results have been obtained combining Zn (which plays a double role as a reactant for hydrogen formation and as a catalyst) and Ni/attapulgite. Yields to GVL higher than 98% have been obtained at 180 °C in the best cases. The best catalytic performance has been related to the presence of tiny Ni particles as nickel crystallites larger than 4 nm were not present in the most efficient catalysts.

Keywords: levulinic acid; γ-valerolactone; hydrogen from water; Zn: Ni; sepiolite; attapulgite

---

1. Introduction

Due to the gradual depletion of oil and restrictions on its use to preserve the environment and reduce climate change effects nowadays, new technologies and methods have been developed to produce fuels and valuable chemicals from renewable resources [1]. New resources have been studied to obtain these chemicals and fuels, especially biomass derivatives. Biomass is a renewable and abundant resource in Nature [2–4] with notable environmental advantages. Many products can be produced from biomass by different methods [5] such as, for example, levulinic acid (LA), which can be easily obtained from carbohydrates or lignocelluloses and it can be transformed into γ-valerolactone (GVL) by hydrogenation through different catalytic routes [6].

Interestingly, GVL has very low toxicity and presents good transport and storage properties. GVL is a colorless liquid with low melting point (−31 °C), open cup flash (96 °C) and high boiling point (207 °C). All these characteristics make GVL an easy and safe to store chemical [7–10]. GVL can be produced cheaply and can be a useful additive to different automotive fuels, reducing undesired emissions while maintaining or barely decreasing performance. GVL can be mixed with gasoline and diesel fuels, leading to a decrease in harmful CO, residual hydrocarbon and particulate matter emissions [11,12]. For example, it has been reported that upon mixing 7% of GVL with diesel fuel the engine performance hardly varies but the smoke produced is reduced by 25% compared to the same diesel fuel without the addition of GVL [13]. Thus, GVL can help combat air pollution and mitigate global warming. In another work, GVL was recommended as a lighter fluid for charcoal and as an illuminating liquid since does not create smoke or odor, producing low levels of VOC emissions [14]. GVL also presents good properties as a green solvent [5,7] and can be a precursor for other valuable chemicals such as alkyl valerates, 4-hydroxypentanol, butanes, 2-methyltetrahydrofuran and adipic acid (nylon) [7–10,15,16].

There are different catalytic alternatives to transform LA into GVL. Thus, this catalytic reaction can be homogeneous or heterogeneous. The heterogeneous pathway is more interesting due to the easier separation of the catalyst and is usually carried out with catalysts containing noble or non-noble metals. Despite the fact non-noble metals are invariably cheaper, they have been less studied because noble metal catalysts give higher conversions at low temperatures, especially those with ruthenium [17–20]. Nevertheless, many studies have used non-noble metals to produce high added value chemicals from biomass [21–28], and nickel sites are some of the most usual active metal sites.

The use of metals deposited on inexpensive supports has been studied for the conversion of LA into GVL [29–32]. In this context, it is of interest to use clay minerals as supports because they are abundant and cheap natural materials. These supports usually present high surface areas and this can lead to a good dispersion of metals on their surface, and as a consequence, this can increase the catalytic activity compared to metals deposited on low surface area supports [33]. Therefore, we have thought that the synthesis of nickel supported on clay minerals can be an interesting option.

The choice of a suitable solvent is also highly important. Unfortunately, there are no solvents without any drawbacks [11,12]. Alcohols, such as isopropanol and 2-butanol, are good options with the additional advantage of promoting the catalytic transfer hydrogenation of levulinic acid into valerolactone. Hydrogen gas solubility in alcohols is slightly higher than in water, but the GVL selectivity is usually higher in water. Alcohols might assist the activation of LA for conversion, but water seems to favor the pathways for selective GVL production. In many cases alcohols are used to mitigate the leaching of metals, which is higher in aqueous media. Other solvents such as dioxane and THF have been also studied as solvents, presenting some advantages in terms of reaction temperature required or extent of metal leaching, but they are toxic and non-environmentally friendly. Overall, water can be considered as the greenest solvent from any viewpoint considered: human health, safety and environment [12].

The catalytic concversion of LA into GVL involves two steps. Firstly, LA has to be hydrogenated to produce 4-hydroxypentanoic acid as a reaction intermediate or dehydrated to form angelica lactone. In the next step the intermediates can form GVL in two possible ways: 4-hydroxypentanoic acid is cyclized and dehydrated or angelica lactone is reduced [11,34–36].

Molecular hydrogen is commonly used in the hydrogenation step [37,38], but it has problems due to its high cost and difficult storage and transport. Formic acid (FA) has been explored as a hydrogen donor, and there are studies about hydrogenation of LA into GVL using this compound with different metallic catalysts [39–41]. FA can be decomposed into H_2 and CO_2 using efficient catalysts, and this hydrogen can be used to transform LA into GVL thus avoiding the need for an external hydrogen supply (Scheme 1). The advantage of using FA for LA hydrogenation lies in the possible cost reduction. Moreover, FA can be formed together with LA when is obtained from sugars.

Another possibility to avoid the use of molecular hydrogen is through the in-situ production of hydrogen from water using metallic catalysts. This catalytic route has been explored in some studies with different non-noble metals for very different applications [42,43]. One of these catalytic routes would involve the use of zinc (Zn) to produce hydrogen from water [44]. Zn has been used in other studies for the transformation of LA into GVL [45]. Then, Zn is able to react with water in order to produce H_2 although in parallel ZnO is also formed. The ZnO formation is, in fact, one of the main problems of using Zn, as it has to be reduced again if we want to re-use the catalyst several times. However, it is possible to reduce the ZnO into Zn in a percentage higher than 90% using solar energy [46–48]. Then, the oxidation of Zn would be less troublesome if solar energy was used to reduce the oxide for a new use [49].

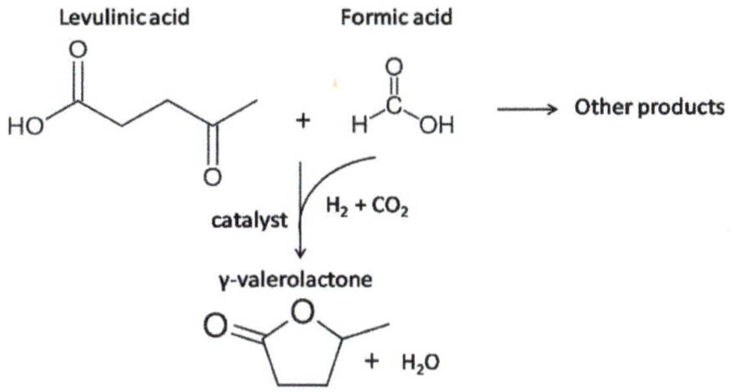

Scheme 1. Reaction of hydrogenation of levulinic acid into γ-valerolactone employing formic acid as hydrogen donor.

Table 1 lists recent studies about Ni catalysts for GVL synthesis using as substrates levulinic acid or its derivatives methyl levulinate (ML) and ethyl levulinate (ET) [50–65]. In many cases, Ni was supported on different materials or mixed with other metals to increase its catalytic activity. Most of these reactions were carried out with molecular hydrogen [50,53–57,59–64], whereas other studies [28,30,51,52] used 2-propanol or formic acid as a hydrogen source [58,64].

Table 1. Overview of Ni catalysts for GVL synthesis from methyl levulinate (ML), ethyl levulinate (EL) and levulinic acid (LA).

Catalyst	Substrate	Solvent	H_2 Source	T (°C)	Time (h)	GVL Yield (%)	Ref.
Ni-Zr-O	ML	Water	Molecular H_2	200 170 150	3 3 3	98 97 95	[50]
Ni-Cu/SBA-15	ML	2-propanol	2-propanol	170	3	87	[30]
RANEY Ni	ML	2-propanol	2-propanol	120	1	94	[51]
RANEY Ni	EL	2-Propanol	2-propanol	80	2	99	[28]
Ni/ZrO$_2$	ML LA	2-propanol 2-propanol	2-propanol 2-propanol	100 120	20 20	94 86	[52]
Ru−Ni/Meso-C	LA	H_2	Molecular H_2	150	2	96	[53]
Ni/NiO	LA	Dioxane	Molecular H_2	120	4,5	99	[54]
Ni/Mg$_2$Al$_2$O$_5$	LA	Dioxane	Molecular H_2	160	1	99	[55]
Ni/Al-LDH	LA	Water	Molecular H_2	200	6	100	[56]
Ni/HZSM-5	LA	Dioxane	Molecular H_2	220	10	93	[57]
Ni-Cu/SiO$_2$	LA	N$_2$	Formic acid	285	-	98	[58]
10NiNb/TiO$_2$	LA	Water	Molecular H_2	275	-	25	[59]
Cu/Ni/Mg/Al	LA	Dioxane	Molecular H_2	140	3	100	[60]
Ni/SiO2	LA	H_2	Molecular H_2	250	0.3	89	[61]
Ni/Al$_2$O$_3$	LA	H_2	Molecular H_2	200	4	92	[62]
Ni/HZSM-5	LA	H_2	Molecular H_2	320	0.5	99	[63]
Ni/SiO$_2$–Al$_2$O$_3$	LA	THF Isopropyl alcohol Water	Molecular H_2 Isopropyl alcohol Formic acid	200	0.5 0.25 10	100 99 70	[64]

In this paper we have studied the use of Ni supported on different clay minerals (two different sepiolites and attapulgite) as a catalyst in the transformation of levulinic acid into γ-valerolactone. Moreover, we have compared the use of different hydrogen sources (formic acid and water as a result of its reaction with Zn) to perform the hydrogenation of LA into GVL at two different temperatures: 120 and 180 °C. The novelty of the present work is focused on two main aspects: (i) molecular hydrogen has not been used and (ii) we have used catalysts containing nickel supported on cheap and abundant materials (extracted from quarries), which have not been previously reported in this reaction.

2. Materials and Methods

2.1. Preparation of Catalysts

Ni was supported on attapulgite, which was collected from Senegal and provided by Sepiolsa (Azuqueca de Henares, Spain). The natural attapulgite used in the present work also contains some other phases, being its percentage composition the following: attapulgite 86 wt.%, smectite: 10 wt.%, quartz: 3 wt.%, dolomite: 1 wt.%. Catalysts were synthesized by a wet impregnation method using ethanol as a solvent. Different Ni concentrations were used by adding the appropriate amount of nickel nitrate ($Ni(NO_3)_2 \cdot 6H_2O$ from Sigma–Aldrich, St. Louis, MO, USA, >98%) and oxalic acid dihydrate ($H_2C_2O_4 \cdot 2H_2O$ also from Sigma–Aldrich) with a molar Ni:oxalic acid ratio of 1:3. Oxalic acid has been used in the synthesis as it is an organic acid with reducing character which is often used in these preparation procedures to decrease the crystallite size of metals and metal oxides such as in the case of nickel, this way increasing the amount of available nickel sites [66,67]. Attapulgite was added to the mixture and placed in a hotplate stirrer at 60 ^{0}C under vigorous stirring until all the solvent was evaporated. These catalysts were labeled as xNi/Atap, being x the Ni content (1, 2 and 5 wt.% were tested).

Two wt.% Ni was selected to synthesize other catalysts using different materials as supports. These supports have been named as Sep and Sep B. Both supports consist of sepiolite collected from Toledo (Spain), and also provided by Sepiolsa. The natural sepiolites used in this work present some other phases. Their composition was: sepiolite 97.0 wt.%, dolomite: 1.9 wt.%, other clays 1.1%. Both supports present the same composition but differ in their crystallinity and surface areas. Thus, the sample called "Sep" is less crystalline than "Sep B". The surface area of "Sep" is 242 $m^2 \cdot g^{-1}$ whilst that of "Sep B" is 381 $m^2 \cdot g^{-1}$. The corresponding supported nickel catalysts were called 2Ni/Sep and 2Ni/SepB, respectively. These sepiolite-based catalysts were prepared by an impregnation method using oxalic acid, as in the case of the Ni/Atap catalysts.

Finally, samples were dried at 100 °C for 12 h and later treated at 500 °C in air for 2 h. Before the reaction and in order to reduce metal oxide precursors, catalysts were heat-treated for 2 h in a flow of H_2 at 400 °C. We decided to reduce in H_2 at 400 °C since at this temperature all the nickel is present as metallic nickel (according to our TEM results). Chemical analysis of 2Ni/Sep, 2Ni/SepB and 2Ni/Atap showed Ni loadings (2.0, 2.1 and 2.1 wt.%, respectively) very close to the theoretical values.

2.2. Characterization Techniques

Catalysts were analyzed by high resolution TEM (HR-TEM), employing a field emission gun TECNAI G2 F20 microscope (FEI Company, Hillsboro, OR, USA) at 200 kV and a JEM 3000F microscope (JEOL, Tokyo, Japan, 300 kV), in order to analyze their structure and morphology. This equipment was also used for energy dispersive X-ray (EDX) and selected area electron diffraction (SAED). Catalyst samples for TEM were sonicated for 20 min in absolute ethanol and deposited on a holey carbon film supported on a copper grid. Then, the copper grid was dried. TEM photographs were used to calculate the average size of the nickel particles. Particle-size histograms were typically constructed by measuring between 80 and 100 particles, depending on the catalyst.

X-ray diffraction (XRD) was used in order to know the crystalline phases of the catalysts and to check if nickel was oxidized or reduced. The apparatus utilized is an Enraf Nonius FR590 sealed tube diffractometer (Bruker, Delft, The Netherlands) equipped with a monochromatic Cu Kα1 source (30 mA and 40 kV).

N_2 adsorption was undertaken at −196 °C, with a Micromeritics ASAP 2460 apparatus (Micromeritics, Norcross, GA, USA). Samples were firstly degassed at 150 °C before reaching vacuum conditions. Total pore volumes were obtained using the adsorbed volume at a relative pressure of 0.97. The specific surface area following the multipoint Brunauer–Emmet-Teller approach (S_{BET}) was obtained at the 0.05 to 0.25 relative pressure range. The pore size distribution was determined employing the Barrett–Joyner–Halenda (BJH) method through the analysis of the N_2 adsorption isotherms.

2.3. Catalytic Tests and Analyses

The catalytic tests for the transformation of LA into GVL were carried out in a 13 mL stainless steel autoclave. The inner steel walls of the autoclave are covered by a Teflon container (made in-house). For the reaction, the autoclave was fed with 3.5 mL of water, 1.24 mmol of LA and 175 mg of catalyst. Three types of reactions were conducted: (i) without any additives, (ii) with the addition of 171.6 mg of Zn and (iii) with the addition of 2.65 mmol of FA. This way the GVL production using different hydrogen sources can be compared. Controls were realized using bare sepiolite, sepiolite B, attapulgite, Zn and FA to test the conversion of LA. Before the reaction, the autoclave was purged with N_2 to minimize the metal oxidation. Later, the autoclave was sealed and introduced in a silicon bath and stirred for 2 h at 800 rpm. This stirring rate was selected in order to ensure a good mixing between the catalyst and the reactants. The reaction was realized at two different temperatures to compare the conversion: 120 and 180 °C. After 2 h, the autoclave was immediately cooled in an ice-bath to stop the reaction. Finally, the dispersion was filtered with an appropriate membrane to obtain the product in order to be analysed.

The analysis of the products has been undertaken as in [21] and [45]. GVL and LA were analyzed by gas chromatography (GC), using a mod. 5890 GC instrument (Hewlett Packard, Palo Alto, CA, USA). This GC has an Agilent HP-1 column (30 m × 0.32 mm × 0.25 μm), a FID detector working at 240 °C, and an injection port at 220 °C. The temperature program for the chromatographic cycle was as follows: (i) 35 °C isothermal for 6 min, (ii) a heating rate of 20 °C min^{-1} from 35 to 230 °C and (iii) 230 °C isothermal for 26 min. Controls without LA and without catalysts were also undertaken and analysed to compare with the other reaction samples. The retention times for γ-valerolactone and levulinic acid are 5.4 and 9.5 min, respectively. Other reaction products were identified using gas chromatography-mass spectrometry (GC-MS 5977A MSD-7890A, Agilent, Santa Clara, CA, USA). The catalytic tests carried out with FA were also analyzed by GC-MS (GC-MS 5977A MSD-7890A, Agilent).

3. Results

Characterization Results

Figure 1 shows the XRD patterns of the Ni catalysts prepared on different supports before (A) and after (B) their reduction. For comparative purposes, the XRD patterns of the supports are also shown.

In the XRD pattern of pure sepiolite, all maxima have been indexed to an orthorhombic symmetry with space group Pnan [68]. Sepiolite is a silicate with fibrous characteristics presenting an ideal formula $[Si_{12}Mg_8O_{30}(OH)_4](H_2O)_4 \cdot 8H_2O$. It has a layered structure made up of large tunnels parallel to the phyllosilicate ribbons (Figure 2A) which are partly occupied by H_2O molecules. These tunnels measure 3.7 × 10.6 Å in cross section.

As it is well known, if sepiolite is heat treated structural modifications due to the loss of water molecules are observed. Preisinger [69] reported that the loss of the bound H_2O leads to a phase change during which the structure is folded by rotation of the phyllosilicate ribbons. These ribbons rotate around an axis through the Si-O-Si corner bonds which link the ribbons (Figure 2B).

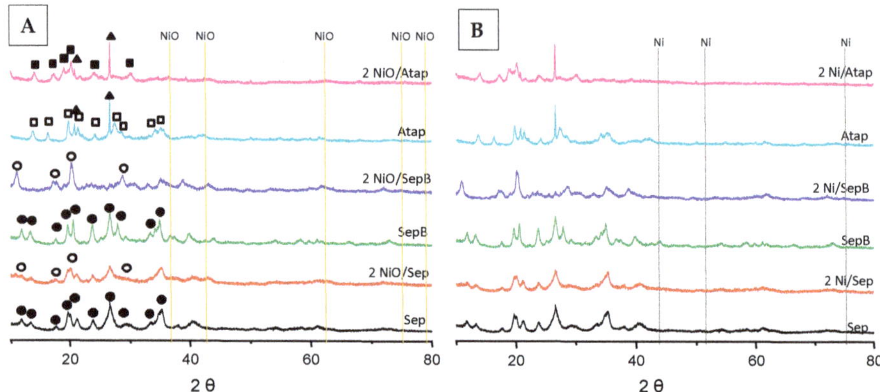

Figure 1. XRD patterns of NiO precursors (**A**) and final Ni catalysts (**B**) and the supports (Sep, SepB and Atap). Symbols: (•) sepiolite, (o) sepiolite-dehydrated, (□) attapulgite, (■) attapulgite-dehydrated, (▲) quartz. – NiO, – Ni.

Figure 2. Schematic representation of sepiolite structure (**A**) before and (**B**) after thermal treatment.

Thus, a phase change takes place at ~600–650 K for the sepiolite heat treated in air. This transition is likely due to the folding of the sepiolite structure when the first moiety of the structural H_2O is lost. This sepiolite "anhydride" is designated as "Sepiol-2H_2O", where "2H_2O" corresponds to the two remaining structural molecules of H_2O.

The Bragg maxima in the XRD pattern of 2Ni/SepB have been indexed on the basis of this structure. The structural change due to the folding becomes irreversible at 570 °C [70]. In this sense, the Ni/Sep catalysts seem to be formed by a mixture of Sep and Sepiol-2H_2O, since the thermal treatment of these phases did not exceed 500 °C.

On the other hand, attapulgite exists in structurally related orthorhombic and monoclinic forms. In the XRD pattern of pure attapulgite (Figure 1), all maxima have been indexed on the basis of an orthorhombic structure. Moreover, there are two maxima that could be assigned to SiO_2 quartz, as indicated with triangles in Figure 1A. Similarly to sepiolite, attapulgite is a fibrous silicate (ideal formula $(Mg_2Al_2)(Si_8)O_{20}(OH)_2 (H_2O)_4$) presenting a layer structure built from large tunnels parallel to the phyllosilicate ribbons which are partly occupied by H_2O molecules. These tunnels measure 3.7 × 6 Å in cross section. In the thermal treatment in air of this clay a structural change to a folded structure also occurs. Indeed, the maxima corresponding to this folded phase, Atap-dehydrated, are observed in the XRD pattern of Ni/Atap samples (Figure 1A).

The XRD patterns of the Ni precursors (Figure 1A) show in the case of the catalysts with attapulgite and sepiolite Sep a pattern very similar to that of the pure supports whereas Sep B was affected by the calcination and a diffractogram with new peaks was observed. We should indicate that the peaks at 2θ = 37.3°, 43.3° and 63.2° are typical of face-centered cubic crystalline structure of nickel oxide (represented by lines in the Figure 1A) and, however, these reflections have not been clearly observed in these precursors although a low intensity peak at 2θ = 43.3° cannot be completely ruled out. This absence of peaks must be due to the fact that the amount of nickel is very low and NiO particles are well dispersed on the support.

In the XRD patterns of the Ni-based catalysts (Figure 1B), it can be observed that catalysts seem to be reduced as the low intensity NiO peak observed in the precursors is now absent. However, it is not possible to see the main peaks of metallic Ni (2θ = 44.5°, 51.1°, 76.1°) likely due to the low concentration of nickel in the catalysts and to the low size of nickel particles.

The textural properties of the Ni catalysts supported on different materials were determined by N_2 adsorption-desorption. BET equation was used to determine the specific surface area and the results are shown in Table 2. All the Ni catalysts displayed reduced specific surface areas compared to the support materials. The surface area of the standard sepiolite (Sep) is 242 $m^2 \cdot g^{-1}$, which slightly decreases after the nickel incorporation. Then, the surface area for 2Ni/Sep is 202 $m^2 \cdot g^{-1}$. The Sep B support reaches a high surface area of 381 $m^2 \cdot g^{-1}$ but drastically decreases until 121 $m^2 \cdot g^{-1}$ in the final Ni-catalyst. This drop is in agreement with the variations observed by XRD in which different patterns are observed in the support and in the final Ni-catalyst. Thus, after the thermal treatment which provokes a folding in the structure, the surface area of the catalysts decreases, in good agreement with the existing literature [70].

Table 2. Textural properties of the Ni-based catalyst supported on different materials.

Sample	S_{BET} ($m^2 \cdot g^{-1}$)	V_T ($cm^3 \cdot g^{-1}$)
Sep	242	0.392
2Ni/Sep	202	0.447
SepB	381	0.619
2Ni/SepB	121	0.579
Atap	216	0.499
2Ni/Atap	75	0.440

In the case of the attapulgite a notable decrease of the surface area was also observed (216 $m^2 \cdot g^{-1}$ in the support vs 75 $m^2 \cdot g^{-1}$ in the final catalyst). Again, this fact could be related to the folding of the structure as a consequence of the thermal treatment, in good agreement with the XRD data, as commented above.

Figure 3A shows the isotherms patterns for the catalysts with 2 wt.% Ni supported on both sepiolites and attapulgite, and their corresponding single supports. According to the IUPAC classification the isotherms of the catalysts and the supports were classified as Type IV [71]. These isotherms patterns are typical for mesoporous adsorbents, therefore showing the mesoporous nature of the sepiolite and attapulgite clays. No big differences were appreciated between the shape of the isotherms corresponding to the supports and those for the final catalysts. However, as mentioned above there are important differences in the amount of N_2 adsorbed.

The pore distribution of the catalysts with Ni content of 2 wt.% is shown in Figure 3B. The original supports present a maximum at ca. 4 nm corresponding to the presence of small mesopores, as typical of this type of clays [72]. However, only in the Ni catalyst on the standard sepiolite (2Ni/Sep catalyst) the maximum at about 4 nm is still visible whereas it is absent in 2Ni/Atap and 2Ni/SepB. This confirms the collapse of the structures of attapulgite and the sepiolite B after the nickel incorporation and the heat-treatments. This data is in agreement with the slight decrease of the surface area in the case of Sep and the drastic drop in the catalysts with attapulgite and the Sep B.

The microstructural characteristics of the supported catalysts were analyzed by high resolution transmission electron microscopy (HRTEM). Figure 4 shows the HRTEM images and corresponding histograms of the 2Ni/Sep, 2Ni/SepB and 2Ni/Attap catalysts.

Clay-support fibers with length varying from 50 nm to several µm and widths ranging between 5 and 20 nm can be clearly observed. The analysis of the nickel-containing particles showed the presence of metal nickel and the absence of NiO particles. Periodical contrasts of 0.21 nm in the images, which correspond to (111) interplanar distances of cubic Ni (inset of Figure 4A), indicate the crystallinity of the Ni nanoparticles. Moreover, when a magnet is placed close to the samples, the particles are attracted by the magnet. This is a typical behaviour of metallic Ni which is ferromagnetic and this would not happen if the particles were NiO, which is antiferromagnetic. Therefore, the presence of Ni nanoparticles can be undoubtedly assumed. These Ni nanoparticles are covering the clay support with similar homogeneity in all the cases and present an average size around 3 nm. In a very small extent, they appear agglomerated in clusters of 6–9 nm in size. Taking into account the particle size distribution histograms (insets in Figure 4A–C), Ni nanoparticles of an average size 3 nm are predominant in all the cases. However, whereas for 2Ni/Sep ca. 25% of the particles are larger than 4 nm, the 2Ni/SepB and 2Ni/Atap catalysts are practically composed only of nanoparticles with sizes less than 4 nm.

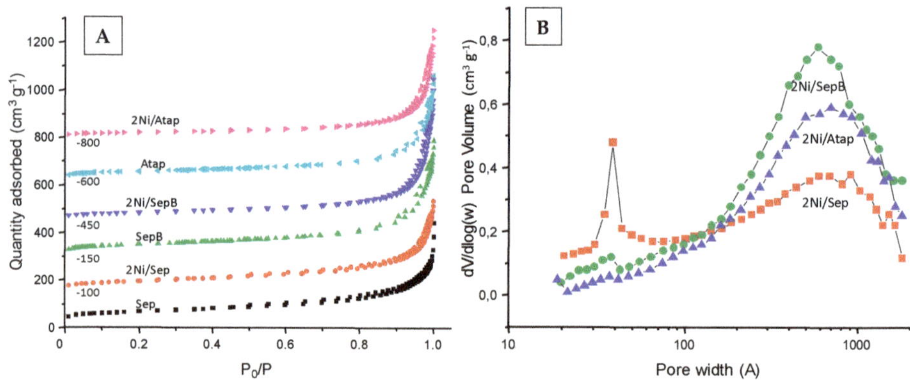

Figure 3. N_2 isotherms patterns for the catalysts with 2 wt.% Ni supported on different materials (**A**) and representative pore distributions (**B**).

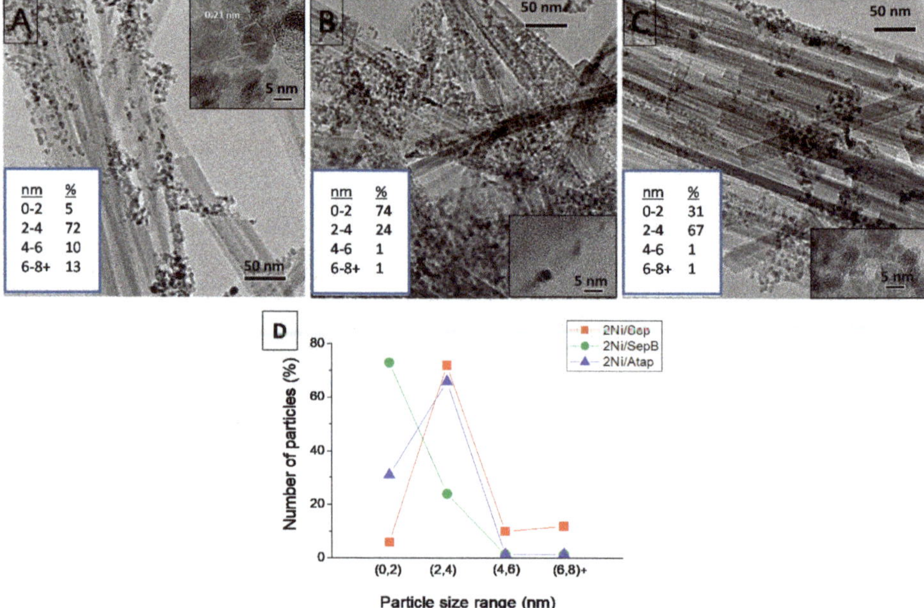

Figure 4. Representative HRTEM images for (**A**) 2Ni/Sep, (**B**) 2Ni/SepB and (**C**) 2Ni/Atap catalysts, and (**D**) distribution of the size of Ni particles in these catalysts.

Initially, different amounts of Ni (1, 2 and 5 wt.%) were supported on attapulgite in order to check the amount of nickel to use with the different supports. These Ni/attapulgite catalysts were tested in the transformation of LA at 180 °C. These preliminary tests were undertaken in the absence of formic acid or Zn. For comparison, data of an unsupported Ni catalyst prepared in the same way (catalyst named as 100Ni) is also included. Figure 5 shows that GVL was produced in all cases. GVL yield was 16% in the unsupported Ni catalyst whereas the blank experiment without catalyst gave no conversion of LA to GVL. It is noteworthy that catalysts with 2 and 5 wt.% Ni obtained yields of ca. 13%, which are very close to that achieved by the catalyst with 100% Ni. The Ni amount of 2 wt.% was selected to prepare the catalysts with the different supports because it has a similar conversion into GVL than 5 wt.%, but its productivity per Ni loading is the highest of all the catalysts tested.

This reaction has been undertaken without an external hydrogen source. However, H_2 proceeds from the water molecules as nickel reacts with water leading to the formation of metal oxide and hydrogen (according to $Ni + H_2O \rightarrow NiO + H_2$). Nickel then acts as both source of hydrogen and catalyst for the AL to GVL transformation.

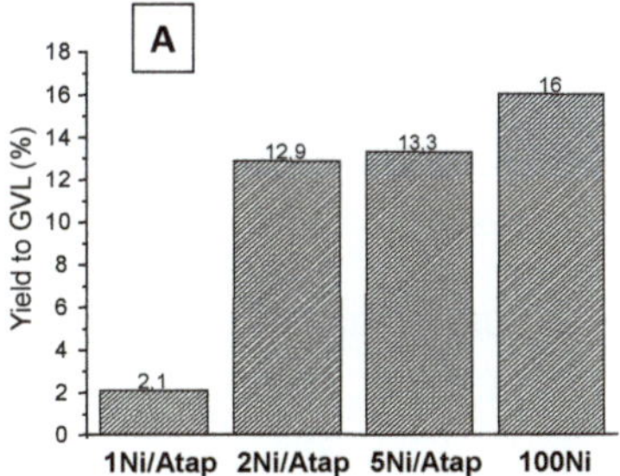
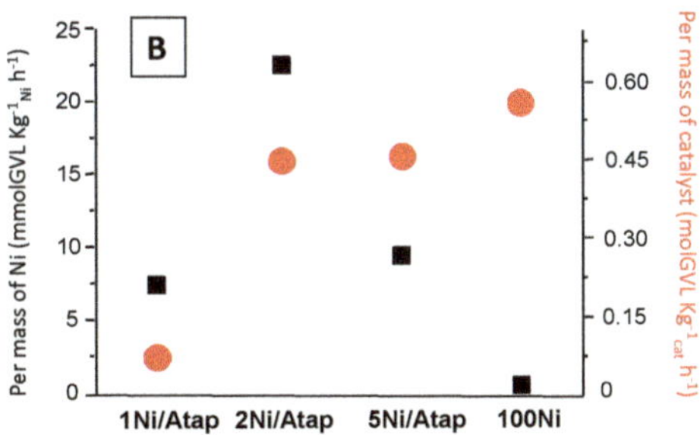

Figure 5. Influence on the Ni-loading in Ni/attapulgite catalysts (**A**) on the conversion of LA into GVL and (**B**) on the GVL productivity per mass of catalyst (•) or per mass of Ni (■). Reaction conditions: T = 180 °C, 3.5 mL of water, 1.24 mmol of LA and 175 mg of catalyst.

Then, a Ni loading of 2 wt.% was selected to be used for the impregnation of the three supports. The first assays were carried out with LA, water and in the absence of another hydrogen source. Then two different assays in the presence of either Zn or formic acid were undertaken in order to produce additional hydrogen (Zn from the reaction with water and formic acid from its decomposition). The experiments were conducted at two different reaction temperatures (120 °C and 180 °C). It must be indicated that in all cases the main reaction product observed has been γ-valerolactone with selectivities over 95%. Then, at the two tested temperatures other compounds were identified by GC-MS as byproducts of the reaction although in low amount. These compounds were 5-methyl 2(3*H*)-furanone and 5-methyl 2(5*H*)-furanone.

Table 3 shows the GVL yield and productivity results obtained by the catalysts with different hydrogen sources at 120 °C and 180 °C. The reactions carried out without hydrogen source with 2Ni/Sep, 2Ni/SepB and 2Ni/Atap show low GVL yield in all cases. At 120 °C the yields to GVL did not exceed 5.1% whereas at 180 °C the yield never reached 15%. As expected, the LA conversion and the GVL formation increase when the reaction temperature increases.

Interestingly, upon adding Zn (that plays mainly the role of reactant as mentioned above) to the nickel catalysts the catalytic activity highly increased. Then at 120 °C in all cases the LA conversion exceeded 70%, being similar with 2Ni/SepB and 2Ni/Atap, giving 82% and 82.7% of GVL yield, respectively. In the case of the 2Ni/sep catalyst the yield to GVL was lower (72.6%). The same trend was observed working at 180 °C. Using 2Ni/SepB and 2Ni/Atap catalysts all the LA reacted, leading to GVL yields higher than 98%. Lower GVL yield (85%) was obtained with 2Ni/Sep catalyst.

Using formic acid in the medium the results were very poor. Thus, the GVL yield results obtained using formic acid as hydrogen source were much worse than those achieved using Zn. In fact, lower yields to GVL were obtained in the presence of formic acid than in its absence. Although formic acid at appropriate temperatures and with a suitable catalyst decomposes into H_2 and CO_2, in the conditions employed in this article the dissociation of acid formic was low and did not help in the production of GVL. Thus, none of the used catalysts gave a GVL yield higher than 7% at the two different tested temperatures. It seems that the presence of formic acid hinders the hydrogen formation from water ($Ni + H_2O \rightarrow NiO + H_2$) originated by the presence of nickel. It is also possible that the negative effect of formic acid is due to the fact that metallic nickel reacts with formic acid, this leading to the formation of hydrogen, and then no active nickel sites are available to undertake the hydrogenation of levulinic acid.

Table 3. Yield of LA into GVL (%) and productivity (mmol GVL/g catalyst/h and mmol GVL/g Ni/h) using Ni on different supports and hydrogen sources.[a]

Catalyst	Additional H_2 Source	GVL Yields		molGVL kg$^{-1}_{catalyst}$ h^{-1}		molGVL kg$^{-1}_{Ni}$ h^{-1}	
		180 °C	120 °C	180 °C	120 °C	180 °C	120 °C
2 Ni/Sep	No	14.6	2.31	0.510	0.082	25.8	4.08
2 Ni/SepB	No	12.0	5.05	0.428	0.180	21.3	8.93
2 Ni/Atap	No	12.9	1.01	0.451	0.040	22.7	1.77
2 Ni/Sep	Zn added	85.2	72.6	3.01	2.57	151	128
2 Ni/SepB	Zn added	>98	82.0	3.54	2.90	177	145
2 Ni/Atap	Zn added	>98	82.7	3.54	2.92	177	146
2 Ni/Sep	Formic acid	5.39	6.50	0.187	0.233	9.52	11.5
2 Ni/SepB	Formic acid	3.02	4.54	0.111	0.156	5.41	8.02
2 Ni/Atap	Formic acid	4.02	6.26	0.142	0.219	7.09	11.1

[a] Reaction conditions: water (3.5 mL), LA (1.24 mmol), 2 wt.% supported metal catalyst (175 mg) reaction temperature (180 °C and 120 °C), time (2 h) and Zn (171.6 mg) or FA (122 mg).

Comparing these three strategies, the use of Zn turns out to be the most effective in transforming LA into GV. Thus, the productivity to GVL is at least one order of magnitude higher when using Zn than when using formic acid or without using additives. Analyses of the gas of the batches demonstrates that in all the experiments done with the Ni-catalysts some hydrogen was formed, being much more abundant in the experiments with Zn. The quantification of the hydrogen has not been possible to be carried out with high accuracy due to experimental limitations.

For comparison a 2 wt.% Ni catalyst supported on silica (Aerosil, surface area 181 $m^2 \cdot g^{-1}$) was prepared in the same way and tested in the LA transformation in the presence of Zn at 180 °C. It was observed that this option also produced high amounts of GVL (yield = 73.7%) but less than comparable catalysts with natural clays as supports. Controls were undertaken in the three conditions mentioned above but in the absence of catalyst. Then, with formic acid or without additives the levulinic acid conversion was negligible. However, in the case of using Zn alone a yield to GVL of 35% was obtained at 180 °C. Then in the experiments undertaken with Ni/catalysts and Zn at 180 °C the catalytic role of both Zn and nickel sites must be considered in order to explain the catalytic performance. In these control experiments hydrogen was detected in the gas-phase in the test using Zn in contrast with the experiment in which additives were not added.

4. Discussion

Ni is an effective non-noble metal capable of producing the conversion of LA into GVL with high efficiency. This way, the use of noble metals is avoided making this process less expensive. In fact, we have observed that it is possible to produce GVL in high yield using small amounts of Ni supported on cheap and highly available supports employing an easy preparation method by impregnation with oxalic acid. The best results have been obtained with 2Ni/SepB and 2Ni/Atap using Zn as a hydrogen source.

In the Ni catalysts supported on clays, nanoparticles cover the support with similar homogeneity and with most particles presenting a size of ca. 3 nm. However, on attapulgite or on the more crystalline sepiolite (Sep B) there are no nickel particles larger than 4 nm, whereas on the standard sepiolite there is an important portion of big particles of 6–9 nm in size. This microstructural feature is probably connected to the better performance of the 2Ni/SepB and 2Ni/Atap catalysts compared to 2Ni/Sep. Then, the higher GVL yields obtained with nickel on attapulgite or on Sep B is likely linked to the lower presence of big nickel particles than on the standard sepiolite. Then, for a fixed nickel loading (2.0–2.1 wt.% in our catalysts) the lower the crystallite size the higher the amount of available active sites is. We want to mention that the catalytic role of the environment around Ni sites cannot be neglected, although the similitude between the support surfaces, lead us to conclude that the main differences are likely due to the different nickel crystallite size. Similarly, in [73,74] it was shown that Ni sites can catalyze hydrogen transfer reactions through a coupling effect, and dispersed Ni nanoparticles present the highest catalytic efficiency in the LA to GVL reaction whereas bare Ni foam could not activate LA. Interestingly, the solvent has a crucial role in the hydrogen transfer reaction because could interact with the metal and react to supply hydrogen.

In order to confirm that metallic Zn had been oxidized to ZnO, samples of Ni-catalysts with Zn were collected after reaction at 120 °C and 180 °C and they were analyzed by XRD. Figure 6A shows the XRD patterns of the 2Ni/sepB catalyst with Zn after reaction at 120 °C and 180 °C. In the sample used at 120 °C, apart from the diffractions of the Ni-catalyst, high intensity peaks corresponding to ZnO (JCPDS: 36-1451) were observed. However, diffractions of metallic Zn (JCPDS: 04-0831) were hardly observed. Then, at 120 °C more than 95% of the initial metallic Zn has been transformed into ZnO in the presence of water with the consequent hydrogen formation. In the sample used at 180 °C, apart from diffractions of ZnO (main crystalline phase detected) and Zn, peaks corresponding to a new Zn-containing phase are observed at 2θ = 12.7, 21.7, 22.1, 25.5, 31.5, 34.0, 38.8 and 49.0° (only cited the most intense). This pattern fits well with that of a Zn-silicate (Zn_2SiO_4 phase, JCPDS: 08-0492). Then, in this case Zn/ZnO reacts with part of the silicon of the sepiolite to form this new phase.

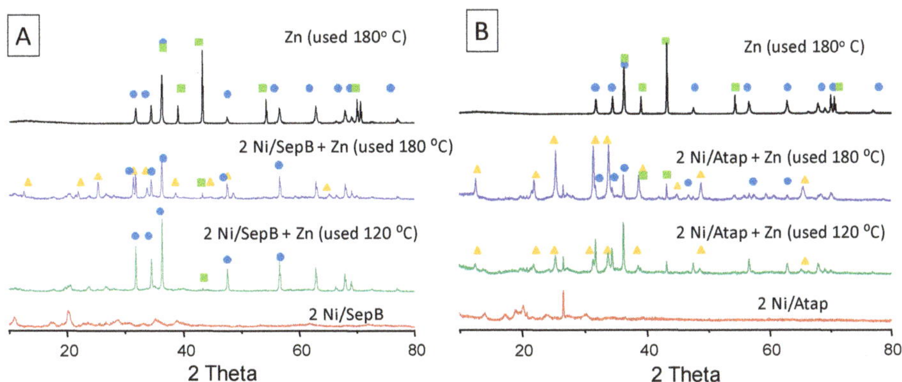

Figure 6. XRD patterns of 2Ni/SepB (**A**) and 2Ni/Atap (**B**): fresh, after use at 120 °C with Zn and after use at 180 °C with Zn. For comparison the XRD pattern of Zn used at 180 °C has been also included. Symbols: (■) Zn^0, (●) ZnO, (▲) Zn_2SiO_4.

Figure 6B shows the XRD patterns of the 2Ni/Atap catalyst with Zn after reaction at 120 °C and 180 °C. In the experiment undertaken at 120 °C apart from the diffractions of the Ni-catalyst, intense peaks of ZnO were detected. The relative intensity of the most intense peaks of ZnO compared to Zn is ca. 5, which is lower than that observed with 2Ni/SepB, in which almost all of the zinc was present as ZnO. Moreover, in this sample the formation of Zn_2SiO_4 phase is already visible at 120 °C in contrast with 2Ni/SepB. This fact could be explained taking into account the presence of some amount of quartz as a secondary phase in the pristine attapulgite, as observed by XRD data and commented above (Figure 1A). In the sample at 180 °C the main reflections correspond to Zn_2SiO_4 phase whereas peaks of lower intensity of Zn and ZnO have also been detected. These results indicate that silicon from attapulgite reacts with Zn/ZnO to form the Zn-silicate phase.

The XRD pattern of the Zn alone sample used at 180 °C shows that ca. 40% of Zn is oxidized into ZnO. This oxidation takes place to a lesser extent than when zinc is together with Ni/SepB or Ni/Atap catalysts (Table 4). It seems that the presence of the nickel catalysts initially accelerates the Zn oxidation to ZnO and then, especially at high temperatures, reacts with a part of the support.

Table 4. Relative intensity of the most intense XRD peaks of Zn, ZnO and Zn_2SiO_4 in used samples.

Experiment/Sample	Reaction Temperature (°C)	ZnO/Zn	Zn_2SiO_4/Zn
Zn alone	180	0.71	-
2Ni/SepB + Zn	120	98	0
2Ni/SepB + Zn	180	8.9	4.0
2Ni/Atap + Zn	120	5.2	2.2
2Ni/Atap + Zn	180	1.5	3.5

A microscopy study of the Zn + 2Ni/SepB sample used at 120 °C shows sepiolite fibers with small patches of nickel together with large particles of ZnO (Figure 7). As indicated above, the length of the sepiolite fibers ranges from 50–100 nm to 4 µm whereas their width varies from 5 to 20 nm. The particles of ZnO are quite heterogeneous in size and shape although most of them are rectangular with typical length of 200–300 nm and width of 80–150 nm. In the initial microscopy study, it was observed that nickel of the fresh Ni/supported catalysts was present as metallic nickel. In Figure 7c we can observe by HRTEM a particle containing nickel deposited on a sepiolite rod. The lighter part of the particle likely corresponds to nickel oxide (d_{hkl} = 0.21 and 0.24 nm, assigned to NiO 200 and NiO 111, respectively) whilst the darkest part probably corresponds to metallic nickel (d_{hkl} = 0.20 and 0.18 nm assigned to Ni 111 and Ni 200, respectively). Then, after the reaction not only is Zn oxidized but also a part of the metallic nickel transforms into NiO. It must be mentioned that the size of Ni/NiO particles in the used catalyst was similar to that of the fresh catalyst and, overall, no enlargement of the particles is apparent. Nevertheless, although most of particles (>95%) were smaller than 6 nm, a couple of large particles (>15 nm) were also observed in the used sample.

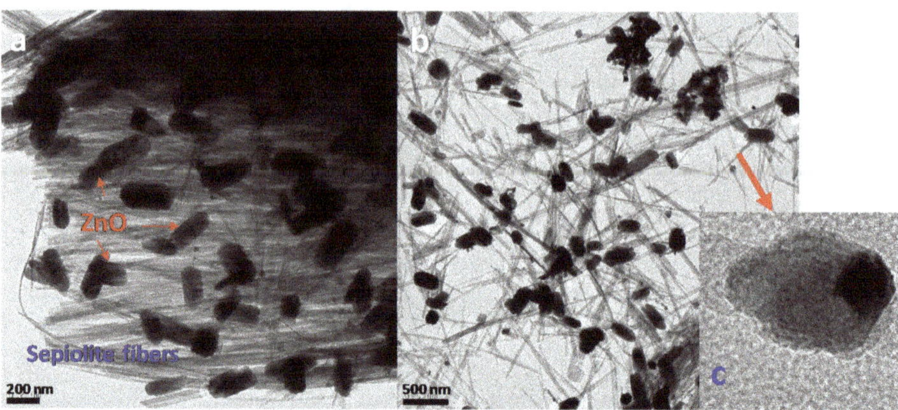

Figure 7. TEM (a,b) and HRTEM (c) images of 2Ni/Sep B with Zn after use at 120 °C.

The reaction using 2Ni/Atap and adding Zn was selected to study the influence of the reaction time in the GVL yield. Two more catalytic tests were carried out for 1 and 6 h at 120 °C (Figure 8).

After 1 h of reaction the yield to GVL was 53.6% and after 2 h the yield to GVL had increased to 82.7%. However, for longer reaction times (6 h) the yield to GVL (84.3%) remained almost identical to that after 2 h. This scarce increase must be due to the fact that after 2 h most of the Zn is already transformed into ZnO (Figure 8B) and then from 2 to 6 h there is hardly any hydrogen available for further reaction. In order to corroborate this, Figure 8B shows the XRD patterns of 2Ni/Atap with Zn after 1, 2 and 6 h of reaction at 120 °C. It can be observed that when the reaction time increases metallic Zn is transformed into ZnO and Zn_2SiO_4. Then, after 1 h there is a high amount of Zn available whereas after 2 h it has drastically decreased. Finally, after 6h there is not any metallic Zn.

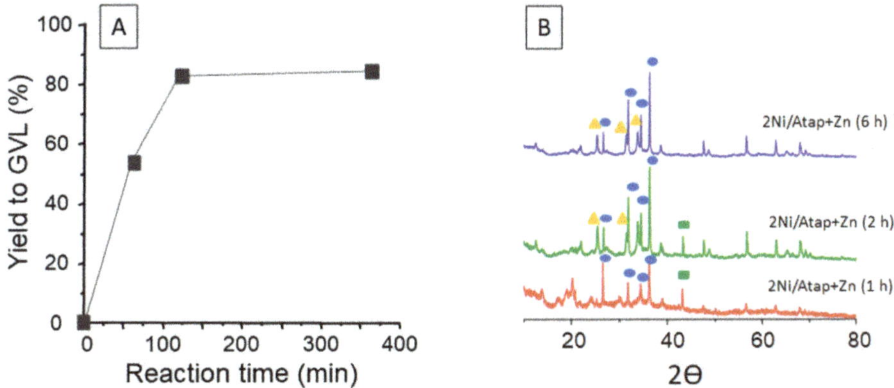

Figure 8. Effect of the reaction time on the yield to GVL (■) using 2Ni/Atap + Zn at 120 °C (**A**) with their corresponding XRD after use (**B**). Symbols: (■) Zn^0, (•) ZnO, (▲) Zn_2SiO_4.

Finally, the recyclability of 2Ni/Atap was studied. After use under standard conditions at 120 °C, the mixture of Ni/Atap + Zn was collected and dried. Then a further catalytic experiment, with a previous reduction, was conducted at 120 °C under standard conditions. Before the second test we had Zn not yet oxidized during the reaction plus metallic Zn obtained from the reduction of the ZnO of the sample used in the first test. As a part of the used mixture could not be collected we assume that the loss of Ni/Atap and Zn is proportional. Then, in the re-testing experiment it was observed a moderate drop in the GVL productivity per mass of catalyst from 2.92 mol GVL $kg^{-1}_{catalyst}$ h^{-1} in the fresh catalyst to 2.29 mol GVL $kg^{-1}_{catalyst}$ h^{-1} in the re-used catalyst. This drop of the productivity could be related to: (i) the formation of the Zn-silicate after use which could modify both the nature of the catalytic sites and the amount of the hydrogen available from the oxidation of Zn and (ii) to the leaching of the nickel active sites. In fact, it can be observed by microscopy that the used catalyst has lower density of nickel nanoparticles than the fresh catalyst. A third re-testing led to a moderate drop in the productivity (2.22 mol GVL $kg^{-1}_{catalyst}$ h^{-1}).

We must indicate that a similar recyclability test at 120 °C but without previous reduction led to a drastic drop of the GVL productivity (from 2.92 mol GVL $kg^{-1}_{catalyst}$ h^{-1} in the fresh catalyst to 0.42 mol GVL $kg^{-1}_{catalyst}$ h^{-1} in the re-used catalyst) probably due to the low hydrogen availability in the reaction medium and maybe to the lack of activation ability of the oxidized nickel.

One of the positive aspects of this catalytic system is the low loss of catalytic activity after use if there is a previous reduction. However, there is a lack of structural stability as during the catalytic reaction the supports react with Zn forming a new phase (Zn-silicates). Interestingly, in spite of this lack of structural stability a drastic deterioration of the catalytic performance has not been observed.

The presence of both metals (Ni and Zn) is essential because Ni is able to transform LA into GVL in the presence of hydrogen and Zn is able to produce hydrogen when it reacts with water. In purity, both nickel sites and Zn can play both roles: (i) produce hydrogen from water and (ii) be catalytically active in the dehydration/hydrogenation of LA to GVL. However, attending to the catalytic results obtained the main role of Zn sites is the supply of hydrogen from its reaction with water whereas the main role of nickel sites is catalytic (hydrogenation/dehydration) (Scheme 2).

Finally, we would like to indicate that the use of metals to produce hydrogen through the reaction with water presents the drawback of the need of a further reduction to be re-used. However, the use of other options such as propanol or formic acid requires new feedings of these compounds, as they are spent during reaction. Therefore, a possible advantage of the approach followed in the present article is that, if the catalytic system is stable enough, it would not be necessary to add fresh metals for the new cycles.

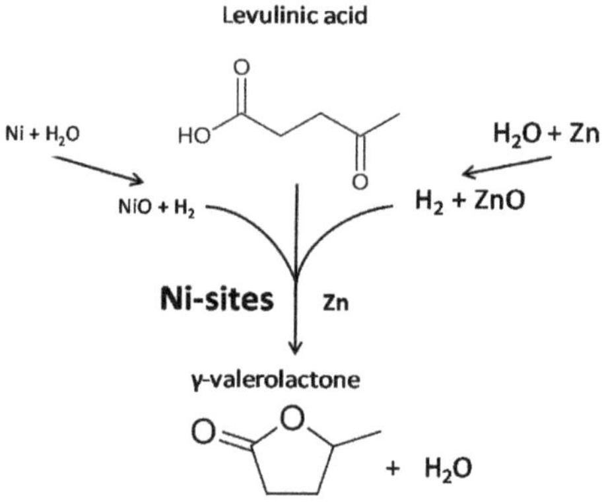

Scheme 2. Reaction of hydrogenation of levulinic acid into γ-valerolactone using the reaction of Zn with water to produce H$_2$ and Ni/sepiolite catalysts.

Overall, the results presented in this article are notorious since GVL yields greater than 98% have been obtained at 180 °C and these are among the best catalytic results reported using other options without feeding molecular hydrogen (Table 1). Now, the next objective is to achieve a further improvement in the stability of the system so that it could be reused without worsening the catalytic performance.

5. Conclusions

Ni-catalysts supported on inexpensive and readily available supports (sepiolites with different crystallinity and attapulgite) were tested in the transformation of levulinic acid into γ-valerolactone in an aqueous medium using three different alternatives without added pressurized H$_2$: (i) no hydrogen source, (ii) adding Zn to the reaction mixture and (iii) adding formic acid. The reactions carried out without any hydrogen source or using formic acid led to low γ-valerolactone yields. The use of Ni supported on attapulgite or on a high surface area sepiolite has led to yields to γ-valerolactone higher than 98% using Zn in the reaction media at 180 °C. Moreover, the structure folding of the clays after thermal treatment seems to provide best yield to GVL. The best performance of nickel on attapulgite or on high surface area sepiolite is likely related to the lower size of nickel particles on these two supports compared to low surface area sepiolite, leading to a higher amount of available active sites. During these experiments, Zn transforms into ZnO and also reacts with the silicon of the supports to form Zn$_2$SiO$_4$, this limiting their reusability. Interestingly, the catalytic performance of the mixture of the Ni-catalyst + Zn can be restored to a large extent.

Author Contributions: Conceptualization, B.S.; methodology, R.S.; formal analysis, F.J.L.; investigation, A.G., R.S., I.Á.-S.; resources, M.P.P., M.L.L.; data curation, A.G., I.V.; writing—original draft preparation, A.G.; writing—review and editing, B.S., F.J.L., I.V.; visualization, F.J.L., I.V.; supervision, B.S.; project administration, M.L.L., I.Á.-S.; funding acquisition, B.S., M.P.P., M.L.L., I.Á.-S. All authors have read and agreed to the published version of the manuscript.

Funding: Authors from UCM thank MINECO (MAT2017-84118-C2-2-R project) and UCM CAI center of EM. Authors from UV thank MINECO (MAT2017-84118-C2-1-R project) an FEDER for funding. A.G. thanks MINECO for the pre-doctoral grant.

Acknowledgments: SCSIE from UV is also acknowledged for characterization of the materials employed and the GC-MS analyses. Said Agouram is acknowledged for assistance in the microscopy experiments. Miranda Sánchez is acknowledged for assistance in the catalytic work.

Conflicts of Interest: The authors declare no conflict of interest.

References

1. Mika, L.; Cséfalvay, E.; Németh, Á. Catalytic conversion of carbohydrates to initial platform chemicals: Chemistry and sustainability. *Renew. Sustain. Energy Rev.* **2017**, *118*, 505–613. [CrossRef]
2. Escobar, J.; Lora, E.; Venturini, O.; Yáñez, E.; Castillo, E.; Almazan, O. Biofuels: Environment, technology and food security. *Renew. Sust. Energ. Rev.* **2009**, *13*, 1275–1287. [CrossRef]
3. Serrano-Ruiz, J.; Sepulveda-Escribano, A. Transformations of biomass-derived platform molecules: From high added-value chemicals to fuels via aqueous-phase processing. *Chem. Soc. Rev.* **2011**, *43*, 5266–5281. [CrossRef] [PubMed]
4. Deng, L.; Li, J.; Lai, D.; Fu, Y.; Guo, Q. Catalytic conversion of biomass-derived carbohydrates into γ-valerolactone without using an external H_2 supply. *Angew. Chem. Int. Ed.* **2009**, *121*, 6651–6654. [CrossRef]
5. Alonso, D.; Wettstein, S.; Dumesic, J. Gamma-valerolactone, a sustainable platform molecule derived from lignocellulosic biomass. *Green Chem.* **2013**, *15*, 584–595. [CrossRef]
6. Orlowski, I.; Douthwaite, M.; Iqbal, S.; Hayward, J.; Davies, T.; Bartley, J.; Miedziak, P.; Hirayama, J.; Morgan, D.; Willock, D.; et al. The hydrogenation of levulinic acid to γ-valerolactone over Cu–ZrO2 catalysts prepared by a pH-gradient methodology. *J. Energy Chem.* **2019**, *36*, 15–24. [CrossRef]
7. Gelosia, M.; Ingles, D.; Pompili, E.; D'Antonio, S.; Cavalaglio, G.; Petrozzi, A.; Coccia, V. Fractionation of Lignocellulosic Residues Coupling Steam Explosion and Organosolv Treatments Using Green Solvent γ-Valerolactone. *Energies* **2017**, *10*, 1264. [CrossRef]
8. Qi, L.; Horváth, I. Catalytic conversion of fructose to γ-valerolactone in γ-valerolactone. *Acs Catal.* **2012**, *2*, 2247–2249. [CrossRef]
9. Zhang, Z. Synthesis of γ-valerolactone from carbohydrates and its applications. *Chem. Sus. Chem.* **2016**, *9*, 156–171. [CrossRef]
10. Horváth, I.; Mehdi, H.; Fábos, V.; Boda, L.; Mika, L. γ-Valerolactone—a sustainable liquid for energy and carbon-based chemicals. *Green Chem.* **2008**, *10*, 238–242. [CrossRef]
11. Dutta, S.; Yu, I.K.M.; Tsang, D.C.W.; Ng, Y.H.; Ok, Y.S.; Sherwood, J.; Clark, J.H. Green synthesis of gamma-valerolactone (GVL) through hydrogenation of biomass-derived levulinic acid using non-noble metal catalysts: A critical review. *Chem. Eng. J.* **2019**, *372*, 992–1006. [CrossRef]
12. Prat, D.; Wells, A.; Sneddon, H.; McElroy, C.R.; Abou-Shehada, S.; Dunn, P.J. CHEM21 selection guide of classical and less classical solvents. *Green Chem.* **2016**, *18*, 288–296. [CrossRef]
13. Bereczky, A.; Lukács, K.; Farkas, M.; Dóbé, S. Effect of γ-Valerolactone Blending on Engine Performance, Combustion Characteristics and Exhaust Emissions in a Diesel Engine. *Nat. Resour.* **2014**, *5*, 177–191.
14. Fábos, V.; Lui, M.Y.; Mui, Y.F.; Wong, Y.Y.; Mika, L.T.; Qi, L.; Cséfalvay, E.; Kovács, V.; Szűcs, T.; Horváth, I.T. Use of Gamma-Valerolactone as an Illuminating Liquid and Lighter Fluid. *Acs Sust. Chem. Eng.* **2015**, *3*, 1899–1904. [CrossRef]
15. Tuck, C.O.; Pérez, E.; Horváth2, I.T.; Sheldon, R.A.; Poliakoff, M. Valorization of Biomass: Deriving More Value from Waste. *Science* **2012**, *337*, 695–699. [CrossRef]
16. Lange, J.P.; Price, R.; Ayoub, P.M.; Louis, J.; Petrus, L.; Clarke, L.; Gosselink, H. Valeric biofuels: A platform of cellulosic transportation fuels. *Angew. Chem. Int. Edit.* **2010**, *49*, 4479–4483. [CrossRef]
17. Ruppert, A.M.; Agulhon, P.; Grams, J.; Wąchała, M.; Wojciechowska, J.; Świerczyński, D.; Cacciaguerra, T.; Tanchoux, N.; Quignard, F. Synthesis of TiO_2–ZrO_2 Mixed Oxides via the Alginate Route: Application in the Ru Catalytic Hydrogenation of Levulinic Acid to Gamma-Valerolactone. *Energies* **2019**, *12*, 4706. [CrossRef]
18. Zhao, D.; Wang, Y.; Delbecq, F.; Len, C. Continuous flow conversion of alkyl levulinates into γ-valerolactone in the presence of Ru/C as catalyst. *Mol. Catal.* **2019**, *475*, 110456. [CrossRef]

19. Van Nguyen, C.; Matsagar, B.; Yeh, J.; Chiang, W.; Wu, K. MIL-53-NH$_2$-derived carbon-Al$_2$O$_3$ composites supported Ru catalyst for effective hydrogenation of levulinic acid to γ-valerolactone under ambient conditions. *Mol. Catal.* **2019**, *475*, 110478. [CrossRef]

20. Xiao, C.; Goh, T.; Qi, Z.; Goes, S.; Brashler, K.; Perez, C.; Huang, W. Conversion of levulinic acid to γ-valerolactone over few-layer graphene-supported ruthenium catalysts. *Acs Catal.* **2015**, *6*, 593–599. [CrossRef]

21. Sanchis, R.; García, T.; Dejoz, A.; Vázquez, I.; Llopis, F.; Solsona, B. Easy method for the transformation of levulinic acid into gamma-valerolactone using a nickel catalyst derived from nanocasted nickel oxide. *Materials* **2019**, *12*, 2918. [CrossRef] [PubMed]

22. Li, J.; Li, M.; Zhang, C.; Liu, C.; Yang, R.; Dong, W. Construction of mesoporous Cu/ZrO$_2$-Al$_2$O$_3$ as a ternary catalyst for efficient synthesis of γ-valerolactone from levulinic acid at low temperature. *J. Catal.* **2020**, *381*, 163–174. [CrossRef]

23. Xie, Z.; Chen, B.; Wu, H.; Liu, M.; Liu, H.; Zhang, J.; Yang, G.; Han, B. Highly efficient hydrogenation of levulinic acid into 2-methyltetrahydrofuran over Ni–Cu/Al$_2$O$_3$–ZrO$_2$ bifunctional catalysts. *Green Chem.* **2019**, *21*, 606–613. [CrossRef]

24. Xu, Q.; Li, X.; Pan, T.; Yu, C.; Deng, J.; Guo, Q.; Fu, Y. Supported copper catalysts for highly efficient hydrogenation of biomass-derived levulinic acid and γ-valerolactone. *Green Chem.* **2016**, *18*, 1287–1294. [CrossRef]

25. Yuan, J.; Li, S.; Yu, L.; Liu, Y.; Cao, Y.; He, H.; Fan, K. Copper-based catalysts for the efficient conversion of carbohydrate biomass into γ-valerolactone in the absence of externally added hydrogen. *Energy Environ. Sci.* **2013**, *6*, 3308–3313. [CrossRef]

26. Tanwongwan, W.; Eiad-ua, A.; Kraithong, W.; Viriya-empikul, N.; Suttisintong, K.; Klamchuen, A.; Kasamechonchung, P.; Khemthong, P.; Faungnawakij, K.; Kuboon, S. Simultaneous activation of copper mixed metal oxide catalysts in alcohols for gamma-valerolactone production from methyl levulinate. *Appl. Catal. A Gen.* **2019**, *579*, 91–98. [CrossRef]

27. Yang, Z.; Huang, Y.; Guo, Q.; Fu, Y. RANEY ®Ni catalyzed transfer hydrogenation of levulinate esters to γ-valerolactone at room temperature. *Chem. Commun.* **2013**, *49*, 5328–5330. [CrossRef] [PubMed]

28. Murugesan, K.; Alshammari, A.; Sohail, M.; Jagadeesh, R. Levulinic acid derived reusable cobalt-nanoparticles-catalyzed sustainable synthesis of γ-valerolactone. *Acs. Sustain. Chem. Eng.* **2019**, *7*, 14756–14764. [CrossRef]

29. Fang, C.; Kuboon, S.; Khemthong, P.; Butburee, T.; Chakthranont, P.; Itthibenchapong, V.; Kasamechonchung, P.; Witoon, T.; Faungnawakij, K. Highly dispersed Ni Cu nanoparticles on SBA-15 for selective hydrogenation of methyl levulinate to γ-valerolactone. *Int. J. Hydrog. Energ.* **2019**, *3*, 272–283. [CrossRef]

30. Li, C.; Ni, X.; Di, X.; Liang, C. Aqueous phase hydrogenation of levulinic acid to γ-valerolactone on supported Ru catalysts prepared by microwave-assisted thermolytic method. *J. Fuel Chem. Technol.* **2018**, *46*, 161–170. [CrossRef]

31. Li, J.; Zhao, L.; Li, J.; Li, M.; Liu, C.; Yang, R.; Dong, W. Highly selective synthesis of γ-valerolactone from levulinic acid at mild conditions catalyzed by boron oxide doped Cu/ZrO2 catalysts. *Appl. Catal. A Gen.* **2019**, *587*, 117244. [CrossRef]

32. Jones, D.; Iqbal, S.; Ishikawa, S.; Reece, C.; Thomas, L.; Miedziak, P.; Morgan, D.; Edwards, J.; Bartley, J.; Willock, D.; et al. The conversion of levulinic acid into γ-valerolactone using Cu–ZrO$_2$ catalysts. *Catal. Sci. Technol.* **2016**, *6*, 6022. [CrossRef]

33. Guerrero-Torres, A.; Jiménez-Gómez, C.; Cecilia, J.; García-Sancho, C.; Franco, F.; Quirante-Sánchez, J.; Maireles-Torres, P. Ni supported on sepiolite catalysts for the hydrogenation of furfural to value-added chemicals: Influence of the synthesis method on the catalytic performance. *Top. Catal.* **2019**, *62*, 535–550. [CrossRef]

34. Jing, Y.; Guo, Y.; Xia, Q.; Liu, X.; Wang, Y. Catalytic production of value-added chemicals and liquid fuels from lignocellulosic biomass. *Chemicals* **2019**, *5*, 2520–2546. [CrossRef]

35. Upare, P.; Lee, J.; Hwang, D.; Halligudi, S.; Hwang, Y.; Chang, J. Selective hydrogenation of levulinic acid to γ-valerolactone over carbon-supported noble metal catalysts. *J. Ind. Eng. Chem.* **2011**, *17*, 287–292. [CrossRef]

36. Putrakumar, B.; Nagaraju, N.; Kumar, V.; Chary, K. Hydrogenation of levulinic acid to γ-valerolactone over copper catalysts supported on γ-Al$_2$O$_3$. *Catal. Today* **2015**, *250*, 209–217. [CrossRef]

37. Manzer, L. Catalytic synthesis of α-methylene-γ-valerolactone: A biomass-derived acrylic monomer. *Appl. Catal. A Gen.* **2004**, *272*, 249–256. [CrossRef]

38. Lange, J.; Vestering, J.; Haan, R. Towards 'bio-based' nylon: Conversion of γ-valerolactone to methyl pentenoate under catalytic distillation conditions. *Chem. Commun.* **2007**, *33*, 3488–3490. [CrossRef]

39. Feng, J.; Gu, X.; Xue, Y.; Han, Y.; Lu, X. Production of γ-valerolactone from levulinic acid over a Ru/C catalyst using formic acid as the sole hydrogen source. *Sci. Total Environ.* **2018**, *633*, 426–432. [CrossRef]

40. Du, X.; He, L.; Zhao, S.; Liu, Y.; Cao, Y.; He, H.; Fan, K. Hydrogen-independent reductive transformation of carbohydrate biomass into γ-valerolactone and pyrrolidone derivatives with supported gold catalysts. *Angew. Chem. Int. Ed.* **2011**, *50*, 7815–7819. [CrossRef]

41. Son, P.; Nishimura, S.; Ebitani, K. Production of γ-valerolactone from biomass-derived compounds using formic acid as a hydrogen source over supported metal catalysts in water solvent. *Rsc Adv.* **2014**, *4*, 10525–10530. [CrossRef]

42. Damma, D.; Smirniotis, P. Recent advances in iron-based high-temperature water-gas shift catalysis for hydrogen production. *Curr. Opin. Chem. Eng.* **2018**, *21*, 103–110. [CrossRef]

43. Mastuli, M.; Kamarulzaman, N.; Kasim, M.; Sivasangar, S.; Saiman, M.; Taufiq-Yap, Y. Catalytic gasification of oil palm frond biomass in supercritical water using MgO supported Ni, Cu and Zn oxides as catalysts for hydrogen production. *Int. J. Hydrogen Energy* **2017**, *42*, 11215–11228. [CrossRef]

44. Jin, F.; Zeng, X.; Liu, J.; Jin, Y.; Wang, L.; Zhong, H.; Yao, G.; Huo, Z. Highly efficient and autocatalytic H_2O dissociation for CO_2 reduction into formic acid with zinc. *Sci. Rep.* **2015**, *4*, 1–8. [CrossRef]

45. Zhong, H.; Li, Q.; Liu, J.; Yao, G.; Wang, J.; Zeng, X.; Huo, Z.; Jin, F. New method for highly efficient conversion of biomass-derived levulinic acid to γ-valerolactone in water without precious metal catalysts. *Acs Sustain. Chem. Eng.* **2017**, *5*, 6517–6523. [CrossRef]

46. Steinfeld, A.; Brack, M.; Meier, A.; Weidenkaff, A.; Wuillemin, D. A solar chemical reactor for co-production of zinc and synthesis gas. *Energy* **1998**, *23*, 803–814. [CrossRef]

47. Osinga, T.; Olalde, G.; Steinfeld, A. Solar carbothermal reduction of ZnO: Shrinking packed-bed reactor modeling and experimental validation. *Ind. Eng. Chem. Res.* **2004**, *43*, 7981–7988. [CrossRef]

48. Haueter, P.; Moeller, S.; Palumbo, R.; Steinfeld, A. The production of zinc by thermal dissociation of zinc oxide—Solar chemical reactor design. *J. Sol. Energy.* **1999**, *67*, 161–167. [CrossRef]

49. Yadav, D.; Banerjee, R. A comparative life cycle energy and carbon emission analysis of the solar carbothermal and hydrometallurgy routes for zinc production. *Appl. Energy.* **2018**, *229*, 577–602. [CrossRef]

50. Sun, M.; Xia, J.; Wang, H.; Liu, X.; Xia, Q.; Wang, Y. An efficient NixZryO catalyst for hydrogenation of bio-derived methyl levulinate to γ-valerolactone in water under low hydrogen pressure. *Appl. Catal. B Environ.* **2018**, *227*, 488–498. [CrossRef]

51. Geboers, J.; Wang, X.; de Carvalho, A.B.; Rinaldi, R. Densification of biorefinery schemes by H-transfer with Raney Ni and 2-propanol: A case study of a potential avenue for valorization of alkyl levulinates to alkyl γ-hydroxypentanoates and γ-valerolactone. *J. Mol. Catal. A Chem.* **2014**, *388–389*, 106–115. [CrossRef]

52. Sakakibara, K.; Endo, K.; Osawa, T. Facile synthesis of γ-valerolactone by transfer hydrogenation of methyl levulinate and levulinic acid over Ni/ZrO2. *Catal. Commun.* **2019**, *125*, 52–55. [CrossRef]

53. Yang, Y.; Gao, G.; Zhang, X.; Li, F. Facile Fabrication of Composition-Tuned Ru–Ni Bimetallics in Ordered Mesoporous Carbon for Levulinic Acid Hydrogenation. *ACS Catal.* **2014**, *4*, 1419–1425. [CrossRef]

54. Song, S.; Yao, S.; Cao, J.; Di, L.; Wu, G.; Guan, N.; Li, L. Heterostructured Ni/NiO composite as a robust catalyst for the hydrogenation of levulinic acid to γ-valerolactone. *Appl. Catal. B Environ.* **2017**, *217*, 115–124. [CrossRef]

55. Jiang, K.; Sheng, D.; Zhang, Z.; Fu, J.; Hou, Z.; Lu, X. Hydrogenation of levulinic acid to γ-valerolactone in dioxane over mixed MgO–Al2O3 supported Ni catalyst. *Catal Today.* **2016**, *274*, 55–59. [CrossRef]

56. Gundekari, S.; Srinivasan, K. In situ generated Ni(0)@boehmite from NiAl-LDH: An efficient catalyst for selective hydrogenation of biomass derived levulinic acid to γ-valerolactone. *Catal. Commun.* **2017**, *102*, 40–43. [CrossRef]

57. Yi, Z.; Hu, D.; Xu, H.; Wu, Z.; Zhang, M.; Yan, K. Metal regulating the highly selective synthesis of gamma-valerolactone and valeric biofuels from biomass-derived levulinic acid. *Fuel* **2020**, *259*, 116208. [CrossRef]

58. Upare, P.P.; Jeong, M.; Hwang, Y.K.; Kim, D.H.; Kim, Y.D.; Hwang, D.W.; Lee, U.-H.; Chang, J. Nickel-promoted copper–silica nanocomposite catalysts for hydrogenation of levulinic acid to lactones using formic acid as a hydrogen feeder. *Appl. Catal. A Gen.* **2015**, *491*, 127–135. [CrossRef]

59. Peddakasu, G.B.; Velisoju, V.K.; Kandula, M.; Gutta, N.; VR Chary, K.; Akula, V. Role of group V elements on the hydrogenation activity of Ni/TiO2 catalyst for the vapour phase conversion of levulinic acid to γ-valerolactone. *Catal. Today* **2019**, *325*, 68–72. [CrossRef]

60. Gupta, S.S.R.; Kantam, M.L. Selective hydrogenation of levulinic acid into γ-valerolactone over Cu/Ni hydrotalcite-derived catalyst. *Catal. Today* **2018**, *309*, 189–194. [CrossRef]

61. Sun, D.; Ohkubo, A.; Asami, K.; Katori, T.; Yamada, Y.; Sato, S. Vapor-phase hydrogenation of levulinic acid and methyl levulinate to γ-valerolactone over non-noble metal-based catalysts. *Mol. Catal.* **2017**, *437*, 105–113. [CrossRef]

62. Hengst, K.; Schubert, M.; Carvalho, H.W.P.; Lu, C.; Kleist, W.; Grunwaldt, J. Synthesis of γ-valerolactone by hydrogenation of levulinic acid over supported nickel catalysts. *Appl. Catal. A Gen.* **2015**, *502*, 18–26. [CrossRef]

63. Popova, M.; Djinović, P.; Ristić, A.; Lazarova, H.; Dražić, G.; Pintar, A.; Balu, A.M.; Novak Tušar, N. Vapor-Phase Hydrogenation of Levulinic Acid to γ-Valerolactone Over Bi-Functional Ni/HZSM-5 Catalyst. *Front. Chem.* **2018**, *6*, 285. [CrossRef] [PubMed]

64. Gundekari, S.; Gundekari, S.; Srinivasan, K.; Srinivasan, K. Screening of Solvents, Hydrogen Source, and Investigation of Reaction Mechanism for the Hydrocyclisation of Levulinic Acid to γ-Valerolactone Using Ni/SiO2–Al2O3 Catalyst. *Catal. Lett.* **2019**, *149*, 215–227. [CrossRef]

65. Yu, Z.; Lu, X.; Liu, C.; Han, Y.; Ji, N. Synthesis of γ-valerolactone from different biomass-derived feedstocks: Recent advances on reaction mechanisms and catalytic systems. *Renew. Sustain. Energy Rev.* **2019**, *112*, 140–157. [CrossRef]

66. Solsona, B.; López Nieto, J.M.; Agouram, S.; Soriano, M.D.; Dejoz, A.; Vázquez, M.I.; Concepción, P. Optimizing Both Catalyst Preparation and Catalytic Behaviour for the Oxidative Dehydrogenation of Ethane of Ni–Sn–O Catalysts. *Top. Catal.* **2016**, *59*, 1564–1572. [CrossRef]

67. Solsona, B.; Concepción, P.; Demicol, B.; Hernández, S.; Delgado, J.J.; Calvino, J.J.; López Nieto, J.M. Selective oxidative dehydrogenation of ethane over SnO2-promoted NiO catalysts. *J. Catal.* **2012**, *295*, 104–114. [CrossRef]

68. Post, J.E.; Bish, D.L.; Heaney, P.J. Synchrotron powder X-ray diffraction study of the structure and dehydration behavior of sepiolite. *Am. Mineral.* **2007**, *92*, 91–97. [CrossRef]

69. Preisinger, A.A. Sepiolite and related compounds: Its stability and application. *Clays Clay Miner.* **1961**, *10*, 365–371. [CrossRef]

70. Ruiz, R.; Del Moral, J.C.; Pesquera, C.; Benito, I.; González, F. Reversible folding in sepiolite: Study by thermal and textural análisis. *Thermochim. Acta* **1996**, *279*, 103–110. [CrossRef]

71. Thommes, M.; Kaneko, K.; Neimark, A.V.; Olivier, J.P.; Rodríguez-Reinoso, F.; Rouquerol, J.; Sing, K.S.W. Physisorption of gases, with special reference to the evaluation of surface area and pore size distribution (IUPAC Technical Report). *Pure Appl. Chem.* **2015**, *87*, 1051–1069. [CrossRef]

72. Tang, Q.; Wang, F.; Tang, M.; Liang, J.; Ren, C. Study on pore distribution and formation rule of sepiolite mineral nanomaterials. *J. Nanomater.* **2012**, *2012*, 382603. [CrossRef]

73. Yan, K.; Liu, Y.; Lu, Y.; Chai, J.; Sun, L. Catalytic application of layered double hydroxide-derived catalysts for the conversion of biomass-derived molecules. *Catal. Sci. Technol.* **2017**, *7*, 1622–1645. [CrossRef]

74. Li, W.; Fan, G.; Yang, L.; Li, F. Highly Efficient Vapor-Phase Hydrogenation of Biomass-Derived Levulinic Acid Over Structured Nanowall-Like Nickel-Based Catalyst. *Chem. Cat. Chem.* **2016**, *8*, 2724–2733. [CrossRef]

Article

Mesoporous Activated Carbon Supported Ru Catalysts to Efficiently Convert Cellulose into Sorbitol by Hydrolytic Hydrogenation

Fatima-Zahra Azar *, M. Ángeles Lillo-Ródenas and M. Carmen Román-Martínez *

MCMA Group, Department of Inorganic Chemistry and Materials Institute, Faculty of Sciences, University of Alicante, Ap. 99, E-03080 Alicante, Spain; mlillo@ua.es
* Correspondence: fz.azar@gmail.com (F.-Z.A.); mcroman@ua.es (M.C.R.-M.)

Received: 17 July 2020; Accepted: 24 August 2020; Published: 26 August 2020

Abstract: Catalysts consisting of Ru nanoparticles (1 wt%), supported on mesoporous activated carbons (ACs), were prepared and used in the one-pot hydrolytic hydrogenation of cellulose to obtain sorbitol. The carbon materials used as supports are a pristine commercial mesoporous AC (named SA), and two samples derived from it by sulfonation or oxidation treatments (named SASu and SAS, respectively). The catalysts have been thoroughly characterized regarding both surface chemistry and porosity, as well as Ru electronic state and particle size. The amount and type of surface functional groups in the carbon materials becomes modified as a result of the Ru incorporation process, while a high mesopore volume is preserved upon functionalization and Ru incorporation. The prepared catalysts have shown to be very active, with cellulose conversion close to 50% and selectivity to sorbitol above 75%. The support functionalization does not lead to an improvement of the catalysts' behavior and, in fact, the Ru/SA catalyst is the most effective one, with about 50% yield to sorbitol, and a very low generation of by-products.

Keywords: Ru nanoparticles; activated carbon; one-pot hydrolytic hydrogenation; cellulose conversion; sorbitol

---

1. Introduction

Concerns on the climate change explain the increasing interest of researchers in low carbon fuels and sustainable energy, which has boosted the investigation on the production of biofuels and chemicals from renewable feedstocks like biomass [1,2]. One of the most attractive approaches is based on the use of non-food biomass, whose conversion into valuable chemicals has recently become the object of study for many researchers dealing with biorefinery processes [1,3–5], and with related reactions such as hydrolysis, pyrolysis, fermentation, dehydration, hydrogenation, etc. A big effort is being made in this field in order to convert cellulose into sugar alcohols. Among these compounds, sorbitol is attractive because of its large number of applications: in pharmacies, food, cosmetics, and as an alternative for biofuel production [6–8]. This polyol can be obtained by two consecutive reactions: hydrolysis of cellulose to produce glucose, followed by glucose hydrogenation to sorbitol, meaning that the process requires two different catalytic functions. In fact, it is usually carried out with catalytic systems consisting of mineral liquid acids (like H_2SO_4 or HCl) as hydrolysis catalysts, and supported metals with activity in hydrogenation reactions [9,10]. However, liquid acids are considered not to be green options, because of their corrosive properties and because they cannot be reused. Stable acidic solids are an interesting alternative because they can be easily recovered from the liquid media [11,12], diminishing its contamination, and they can, potentially, be reused. Because of that, they can be considered promising catalysts that could replace liquid acids to make the processes greener.

This is the context of this proposal, which focuses on the use of heterogeneous bifunctional catalysts to obtain sorbitol from cellulose by the combined reaction known as hydrolytic hydrogenation. A suitable heterogeneous catalyst must be composed of a solid acid and supported metal particles to enhance, respectively, cellulose hydrolysis and the hydrogenation of hydrolysis products. Examples of recent works dealing with this topic are shown in references [13–16].

Considering the simultaneous need of acidic and metallic active sites for this approach, it is necessary to focus on a solid able to bear both acidic functional groups and metal nanoparticles on its surface. Carbon materials can easily be such a solid because both the surface area and the surface chemistry can be tuned in order to suit their properties to the desired application. In particular, acidic sites can be created without detriment of the textural properties. Besides, it has been fully recognized that they have outstanding properties as catalyst supports in many reactions [17,18].

Among the metals active for the target reaction, Ru and Ni have proven to be very interesting [10,19,20]. In fact, this combination, catalysts based on carbon supported Ru and Ni metallic particles has been studied before, showing outstanding results. For example, Komanoya et al. reported 68% sorbitol yield with a Ru/active carbon (AC) catalyst and a mix milling process [21], and Deng et al. also reported 69% sorbitol yield with Ru supported on carbon nanotubes (CNT), but using a concentrated H_3PO_4 solution for cellulose pretreatment [22]. More recently, Ribeiro et al. reported 86% sorbitol selectivity, also using mix milling and Ru-Ni bimetallic catalysts supported on AC and CNT [23].

The present work deals with the preparation and characterization of Ru catalysts supported on activated carbons for the hydrolytic hydrogenation of cellulose. The pristine selected activated carbon is a commercial one, named SA, interesting mainly for its developed mesoporosity. Two further supports, named SASu and SAS, have been prepared from SA by sulfonation and oxidation treatments, respectively, with the purpose of increasing the carbon acidity. The relatively high mesoporosity of these carbon supported Ru catalysts is expected to facilitate the access of cellulose to the catalyst's surface, thus enhancing the interaction with the active sites, particularly acidic sites that catalyze cellulose hydrolysis. According to the research work of Chung et al. [24], the adsorption and transformation of β-(1→4)-glucans can be enhanced in the pores of a mesoporous carbon material, where the long cellulose chains can be more easily hydrolyzed to glucose monomers.

2. Materials and Methods

2.1. Catalysts Preparation and Characterization

The commercial powder activated carbon SA-30 from MeadWestvaco (USA) was used for this study, with abbreviated name SA. It was submitted to the following chemical treatments: A- 1 M H_2SO_4, and B- $(NH_4)_2S_2O_8$ saturated solution in 1 M H_2SO_4. In both treatments, the 1 g carbon/10 mL solution mixture was stirred for 24 h at room temperature. Afterwards, the solid was recovered by filtration and washed several times with distilled water (up to the total elimination of sulfates in the filtrate (determined by $BaCl_2$ testing)). The samples resulting from these treatments are named as SASu and SAS, respectively.

Ru nanoparticles were supported on the carbon samples (SA, SASu and SAS) as follows: each degasified carbon sample (150 °C, 4 h, vacuum) was put in contact with a $RuCl_3$ aqueous solution (1 g carbon/10 mL solution) of the appropriate concentration to obtain 1 wt% Ru loading, and kept under stirring for 14 h at room temperature. Afterwards, the mixture was stirred in an ultrasound bath for 3 h. Finally, the solvent was removed at 60 °C for 10 h, and then it was dried at 110 °C for 24 h. Before being used in a catalytic activity test, the Ru containing samples, named Ru/SA, Ru/SAS and Ru/SASu, were submitted to a reduction treatment under H_2 flow (80 mL/min), at 250 °C for 4 h.

Preparation conditions have been selected after previous, still unpublished, works.

The textural properties of the original and Ru containing carbon materials were determined by N_2 adsorption-desorption at −196 °C (Autosorb-6B, Quantachrome). The specific surface area (S_{BET}) and the total micropore volume (V_{micro}) were determined from N_2 adsorption data by applying the Brunauer–Emmett–Teller (BET) and the Dubinin–Radushkevich (DR) equations, respectively. To estimate the mesopore volume (V_{meso}), the difference of the volume of N_2 adsorbed as liquid at $P/P_0 = 0.9$ and $P/P_0 = 0.2$ was calculated [25,26].

The surface chemistry of the prepared catalysts was studied by temperature programmed desorption (TPD) experiments in the following conditions: the sample (9–12 mg) was heated at 20 °C/min in He flow (100 mL/min) up to 900 °C. The equipment used was a thermobalance (TA-SDT Q600) coupled to a mass spectrometer (Thermostar, Balzers), allowing the simultaneous record of weight loss and the analysis of evolved gases (CO_2, CO, and H_2O) during the experiment.

Transmission electron microscopy (TEM, JEOL JEM-2010), with the Infinity Analyze software for image analysis, was used to analyze the size and distribution of the supported Ru particles.

X-ray photoelectron spectroscopy (XPS, VG Microtech Multilab ESCA-3000 spectrometer) was used to characterize the surface chemical composition and the electronic state of Ru in the prepared catalysts.

2.2. Cellulose Hydrolytic Hydrogenation

The commercial Avicel microcrystalline cellulose (99%, Sigma Aldrich) was ball-milled (agate balls/cellulose weight ratio of 3500 rpm, with reverse rotation every 60 min, for 7 h). Catalytic tests were carried out in a 50 mL stainless steel Parr reactor (Model 4792) lined with a Teflon container, and equipped with a manometer and a thermocouple (see Figure S1 (S accounts for Supplementary Material)), as follows: 500 mg milled cellulose, 125 mg catalyst and 25 mL distilled water, together with a magnetic rod, were introduced in the reactor. After closing the reactor, it was purged several times to remove air, and then, it was filled with H_2 and heated, under stirring, to reach the reaction conditions of 50 bar and 190 °C. The reaction time was 3 h in all catalytic tests. Reaction conditions have been selected after previous, still unpublished, works. To determine the reaction progress, the solid and liquid phases were separated by filtration, after cooling down, and then, the liquid phase was analyzed by high performance liquid chromatography (HPLC, 1200 infinity Agilent Technologies Hi-Plex Ca (Duo), 300 × 6.5 mm. An example of the obtained spectra is shown in Figure S2. The remaining solid (catalyst and unreacted cellulose) was dried and weighted to calculate the cellulose conversion.

The products yield was calculated from HPLC results, as indicated in Equation (1):

$$\text{Yield (to A)} = ((\text{Mol of A})/(\text{Mol of charged cellulose})) \times 100 \qquad (1)$$

Conversion was calculated as shown in Equation (2):

$$\text{Conversion} = (1 - (\text{Weight of unreacted cellulose})/(\text{Weight of charged cellulose})) \times 100 \qquad (2)$$

Moreover, selectivity was calculated in terms of conversion and yield as follows (Equation (3)):

$$\text{Selectivity} = (\text{Yield/Conversion}) \times 100 \qquad (3)$$

3. Results

3.1. Textural Properties

The −196 °C N_2 adsorption-desorption isotherms are presented in Figure S3. They are type IV according to the IUPAC classification [27]. The relatively high adsorption at low relative pressure and the steep slope indicates that these materials contain significant volumes of both micro and mesopores.

The textural parameters calculated from the isotherms data are shown in Table 1. Carbon SA presents high surface area and pore volume, with a similar proportion of micro and mesopore volumes. Treatment A (1 M H_2SO_4), that leads to sample SASu, results in a slight modification of the porous structure of the SA carbon, while treatment B (($NH_4)_2S_2O_8$ saturated solution in 1 M H_2SO_4), leading to sample SAS, produces a significant decrease of the adsorption capacity of the original carbon material. Such an effect can be due either to the destruction of pores, or to some blockage of the porosity by the abundant surface oxygen groups. This was previously reported for other carbon materials submitted to a similar treatment [28].

Table 1. Textural parameters determined from N_2 adsorption isotherms (at $-196\ °C$).

Sample	S_{BET} [a] (m^2g^{-1})	V_{micro} [b] (cm^3g^{-1})	V_{meso} [c] (cm^3g^{-1})
SA	1464	0.73	0.74
SASu	1522	0.78	0.60
SAS	1274	0.65	0.48
Ru/SA	1416	0.71	0.66
Ru/SASu	1406	0.68	0.63
Ru/SAS	1218	0.60	0.47

[a] BET surface area and [b] micropore volume, determined by applying the Brunauer–Emmett–Teller (BET) and the Dubinin–Radushkevich (DR) equations to the N_2 adsorption data, respectively; [c] mesopore volume estimated by the difference of the volume of N_2 adsorbed as liquid at $P/P_0 = 0.9$ and $P/P_0 = 0.2$.

The incorporation of Ru produces only a slight decrease of the surface area and porosity of SA, SASu and SAS carbons, indicating that the supported Ru nanoparticles produce almost no blockage of the carbon porosity. Thus, comparing the Ru-containing catalysts, it can be observed that Ru/SA and Ru/SASu show a similar porosity, somewhat higher than that of Ru/SAS.

3.2. Surface Chemistry

Both original and Ru-containing carbon materials were characterized by TPD. Table 2 includes the quantification of the obtained TPD profiles. The TPD-CO_2 and TPD-CO curves of the three Ru-containing catalysts are shown in Figure 1, while those corresponding to the carbon materials can be seen in Figure S4.

The evolution profiles of CO_2 and CO, produced by the decomposition of oxygen functional groups (OFG), can give information about the nature of such OFG. Those that decompose at lower temperature, mainly as CO_2, have acidic character (carboxylic, anhydrides and lactones), and those that decompose as CO are weakly acidic, like phenol type groups, or basic, like carbonyl and quinone type groups. Based on the reported temperature intervals for the thermal decomposition of each type of group [29–32], Figure 1 shows their approximate distribution in the TPD profiles (see the figure caption for the identification of oxygen functional groups).

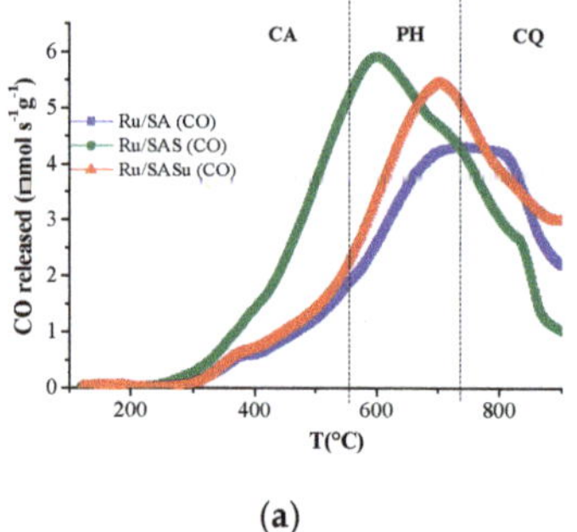

(a)

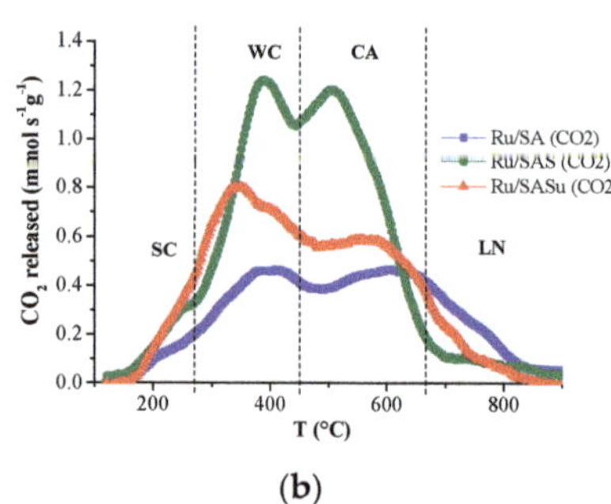

(b)

Figure 1. CO (**a**) and CO_2 (**b**) TPD profiles for the Ru containing samples (SC: strong carboxylic acid, WA: weak carboxylic acid, CA: carboxylic anhydride acid, PH: phenol, CQ: Carbonyl and Quinone, LC: Lactone).

Table 2. Quantification of evolved CO_2 and CO in TPD experiments and the corresponding calculated O wt%.

Sample	CO_2 (mmolg^{-1})	CO (mmolg^{-1})	$CO + CO_2$ (mmolg^{-1})	O [a] (wt %)	Acidic OFG [b] (mmol/g)
SA	0.5	2.8	3.3	5.7	1.9
Ru/SA	0.7	4.5	5.2	8.6	2.9
SASu	0.7	3.6	4.3	7.4	2.5
Ru/SASu	0.9	5.3	6.2	10.2	4.0
SAS	2.2	6.3	8.5	14.6	7.2
Ru/SAS	1.2	6.0	7.2	11.8	4.5

[a] determined from data of (mmol O)/g calculated as (mmol CO)/g + (2 mmol CO_2)/g. Moreover,
[b] determined from deconvolution data, as the sum of quantification corresponding to SC, WC, CA, PH and LN functional groups.

Pristine SA carbon has an intermediate amount of OFG compared to other activated carbon materials that have been reported in the literature, but the large content of groups that evolve as CO can be pointed out. It can be observed that treatment A renders a moderated amount of surface oxygen functional groups (sample SASu), while treatment B resulted, as expected, in an important development of surface functional groups (sample SAS). Thus, the sum of evolved CO_2 and CO is, in SAS, more than double than in SASu.

A comparison of the TPD results corresponding to the carbon materials and the carbon supported Ru catalysts (Table 2 and Figure S4) shows that the amount of CO_2 + CO released from Ru/SA and Ru/SASu is larger than the amount released from the respective supports. This means that the preparation process of the supported Ru nanoparticles (impregnation and reduction treatment) leads to a certain restructuration of the carbon surface chemistry. This likely involves the creation of new groups and the transformation of previously existing ones [33,34]. Besides, the Ru nanoparticles can catalyze the decomposition and/or transformation of OFG during the TPD experiment and, thus, the TPD profiles are much more difficult to interpret because of that. In contrast, the evolution of CO_2 and CO decreases after Ru incorporation in the SAS carbon. In this case, the net balance of the effect of the catalyst preparation steps (impregnation and reduction) is the loss of a certain amount of OFGs from the surface of the highly oxidized SAS support.

Deconvolution of the TPD curves, according to the criteria from previous works [32,35,36], was performed in order to better identify the nature of functional groups, and to try to quantify all of them (the deconvoluted curves can be seen in Figure S5). As indicated above, the main OFG are carboxylic acid (weakly (WC) and strongly (SC) acidic), carboxylic anhydride (CA), phenol (PH) and lactone (LN), with acidic character, and carbonyl and quinone (CQ), with basic character.

The calculated amount of each type of functional groups in the carbon materials and in the Ru catalysts is shown as bar diagrams in Figure 2. The experimental error in these data is below 5%, estimated from the precision of data usually obtained in TPD experiments performed with the device used in this work.

It can be observed that, upon treatment A (sample SASu), basic CQ groups are mainly developed, followed by carboxylic anhydride (CA) groups, whereas treatment B (sample SAS) leads to an extensive development of carboxylic anhydride (CA) and phenol type (PH) groups. Regarding the effect of Ru incorporation, and compared to the respective supports, data of Figure 2 show that: in the case of the Ru/SA sample, the increase in the amount of CQ groups is very important and the increase in the amount of CA and PH groups is also relevant; in Ru/SASu, the content of all types of OFG, excepting SC and WC increases, and in Ru/SAS, in spite of the general decrease of the OFG amount, the CQ and LN groups content increases.

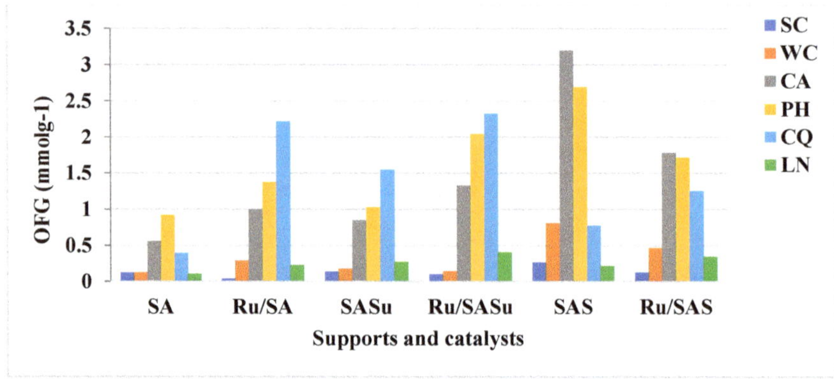

Figure 2. OFG distribution determined by deconvolution of the TPD spectra (SC: Strong acidic carboxyl, WC: Weak acidic carboxyl, CA: Carboxylic anhydride, CQ: Carbonyl or Quinone, PH: Phenol, LN: Lactone).

According to these data, the amount of acidic sites present in the studied samples has been estimated as sum of the amount of SC, WC, CA, PH and LN groups, and has been included in Table 2. This table shows that for both, the carbon supports and the analogous metal containing catalysts, the acidic groups content follows the relative order: SA < SASu < SAS.

3.3. XPS and TEM Analysis

XPS analysis of the three Ru containing catalysts was carried out to determine the electronic state of the supported Ru species, to identify potential sulfur species present, and to analyze the oxygen species, to complete the information obtained by TPD. Figure 3 shows the Ru 3p XPS spectra of the three studied catalysts.

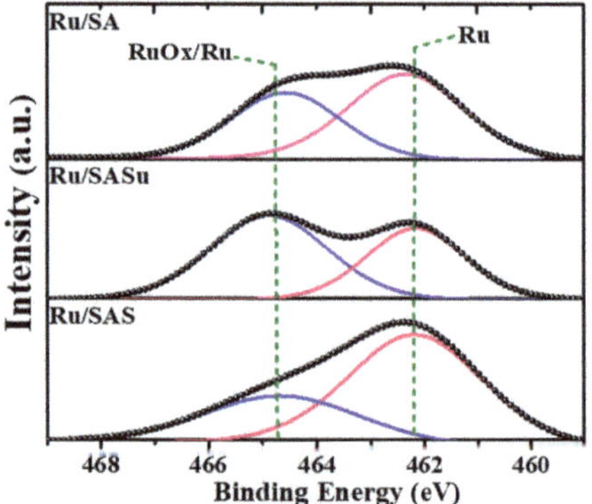

Figure 3. Ru 3p XPS spectra of the three studied catalysts.

The Ru 3p fitted spectra of the three Ru catalysts reveal the presence of metallic Ru (B.E = 462.3 eV) [37], and oxidized Ru species (B.E. = 464.7 eV) [38,39]. This means that the reduced catalysts become partially oxidized upon exposition to air, being the proportion of metallic Ru respect to total Ru close to 50% (56%, 41% and 67% in Ru/SA, Ru/SASu and Ru/SAS catalysts, respectively; the differences cannot be considered as fully significant, because the time of air exposure after the reduction treatment was not exactly the same for the three catalysts). In any case, the oxidized species will probably be reduced again to the metallic state during the reaction, because of the presence of high-pressure hydrogen at 190 °C. The total amounts of Ru determined by XPS are 1.13wt%, 1.02wt% and 1.52 wt% in Ru/SA, Ru/SASu and Ru/SAS, respectively, which is in line with the nominal metal loading of the catalysts.

Figure 4 shows the O 1s XPS spectra deconvoluted according to the B.E. assignments reported in the literature [30,40,41]. It can be seen that they show mainly four signals corresponding to the C=O bond in quinone groups (about 531.0 ± 0.4 eV), C=O or –OH related to lactone or hydroxyl groups (532.0 ± 0.2 eV), C–OH attributed to phenol groups (at 533.0 ± 0.3 eV) and –COOH (534.0 ± 0.5 eV), assigned to carboxylic groups.

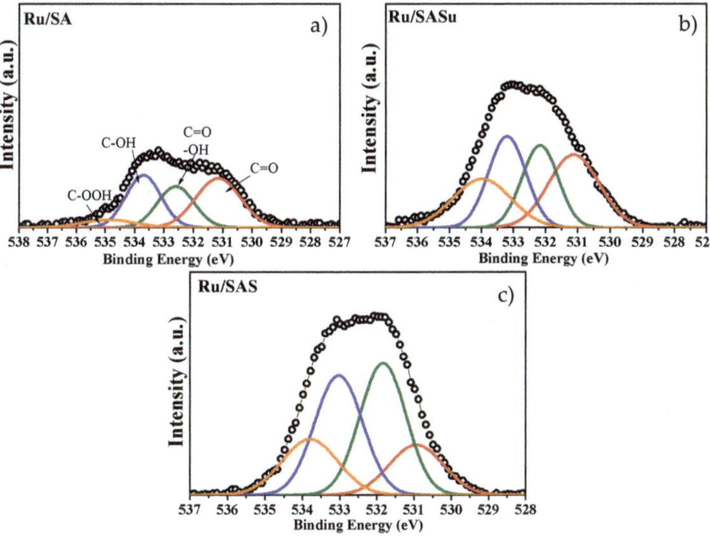

Figure 4. O 1s XPS spectra of: (**a**) Ru/SA, (**b**) Ru/SASu and (**c**) Ru/SAS. The code color indicated is the same in Figures (**a–c**).

The quantification of the O1s spectra, as oxygen wt %, in all the mentioned oxygen-containing surface species, is presented in Table 3.

Table 3. Distribution of OFG as O wt% determined from the deconvolution of the O 1s XPS peaks.

	O wt %					
Peak	1	2	3	4	O in acidic OFG [a]	**Acidic OFG [b] (mmol/g)**
B.E. (eV)	531 ± 0.4	532 ± 0.2	533 ± 0.3	534 ± 0.5		
Sample	C=O	C=O or OH	C-OH	C-OOH		
Ru/SA	3.32	2.26	2.77	0.62	3.39	1.93
Ru/SASu	3.63	2.91	3.36	2.75	6.11	2.99
Ru/SAS	3.34	7.51	7.05	3.83	10.88	5.60

[a] sum of O wt% in C-OH and C-OOH groups (from data in columns 3 and 4). [b] calculated from O wt. % in C-OH and C-OOH groups (from data in columns 3 and 4), taking into account that column 4 corresponds to functional groups with two oxygen atoms.

Considering that the main acidic groups determined by XPS are carboxylic and phenol type ones, the O wt% related to these groups (columns 3 and 4 in Table 3) has been summed up and assimilated to the amount of O in acidic oxygen functional groups. Data show that it increases from Ru/SA to Ru/SAS, in agreement with the TPD data. For a more proper comparison of XPS and TPD data, the amount of oxygen in acidic groups has been calculated as mmol/g of acidic groups (taking into account that column 4 corresponds to functional groups with two oxygen atoms). The calculated values are included in Table 3. Compared with analogous data of Table 2, it can be concluded that the results of both techniques are in line (the differences can be related with inherent differences in the fundamentals of analysis of the two techniques).

The S 2p XPS spectra of the Ru/SASu and Ru/SAS catalysts reveal the presence of sulfonic groups (peak at ~168.5 ± 0.1 eV) [42]. However, those spectra present a lot of noise, which makes the quantification of sulfur species quite imprecise.

Figure 5a–f show TEM images obtained for the Ru catalysts. It must be mentioned that although, in general, well dispersed Ru nanoparticles have been observed (Figure 5a–c), some agglomeration of particles has also been found (Figure 5d–f), particularly in the case of the catalyst prepared with the most oxidized support. The particle size distribution has been plotted as bar diagrams in Figure 5g–i, being 1.3, 1.4 and 1.6 nm the average particle sizes in Ru/SA, Ru/SASu and Ru/SAS catalysts, respectively. Figure S6 shows a TEM image of the original SA activated carbon.

In general, TEM data indicate that the impregnation method used was successful for the formation of small Ru nanoparticles (average size lower than 2 nm) on SA, SASu and SAS carbon materials, and it seems that the particle size slightly increases with the support oxidation. This can be related to the anchorage of the Ru precursor species on the OFG, and the effect of their partial decomposition during the reduction heat treatment, which could lead to some metal sintering [43–45].

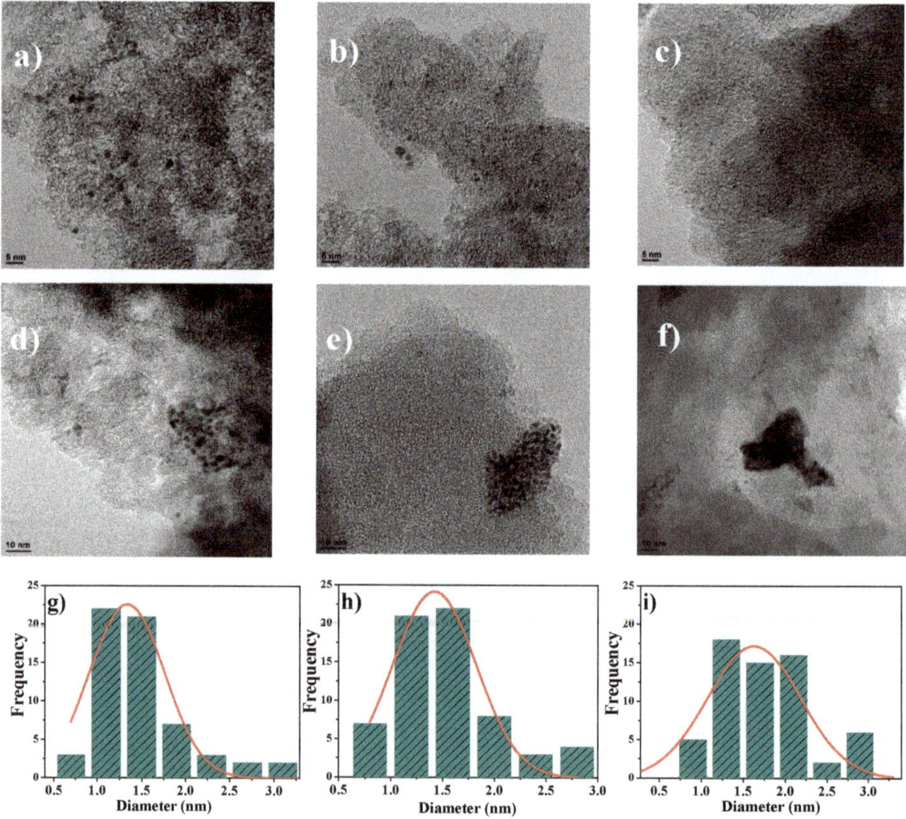

Figure 5. TEM images and Ru particle size distribution of catalysts: Ru/SA (**a,d,g**); Ru/SASu (**b,e,h**); and Ru/SAS (**c,f,i**).

3.4. Catalytic Conversion of Cellulose

Table 4 shows the obtained cellulose conversion and yield of the main reaction products values. Other products (not shown here) have also been obtained in a very low amount, and most of them could not be identified. Figure 6 shows a simplified scheme of the reaction pathway from cellulose to the products presented in Table 4.

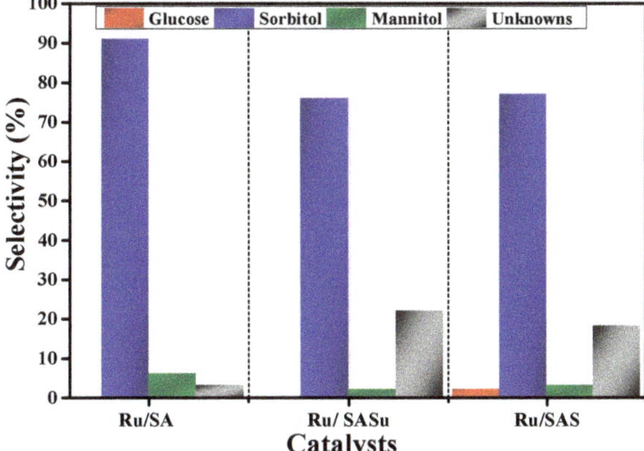

Figure 6. Scheme of the reactions and products considered in the present work.

Table 4. Cellulose conversion and products yield.

Catalyst	Conversion %	Yield %			
		Glucose	HMF	Sorbitol	Mannitol
Blank	35	13	10	-	-
Ru/SA	52	-	-	48	3
Ru/SASu	46	-	-	35	1
Ru/SAS	43	1	-	33	1

Reaction conditions: 500 mg cellulose, 125 mg catalyst, 25 mL water, 190 °C, 50 bar, 3 h.

In the blank test (without catalyst), cellulose conversion was 35%, and small amounts of glucose and hydroxymethylfurfural (HMF) were formed (Table 4). In the tests carried out with the carbon supported Ru catalysts, a noticeable increase of cellulose conversion compared with the blank experiment and a high sorbitol yield have been achieved. The experimental error in these data is below 5%. It has been estimated from the deviation determined when catalytic activity tests have been repeated.

Selectivity data have been plotted in Figure 7.

Figure 7. Selectivity data of the three studied catalysts.

Data from Table 4 show that, among the Ru containing catalysts, Ru/SA is the best performing one in terms of cellulose conversion and sorbitol yield. As shown in Figure 7, the three catalysts are quite selective to sorbitol (selectivity above 75%), being Ru/SA highly selective (91%). It can be pointed out that catalysts Ru/SASu and Ru/SAS produce a significant amount of by-products.

Catalyst Ru/SA was tested for recyclability. After the first run, the catalyst and the unreacted cellulose were recovered from the reaction media by filtration, and after washing and drying, they were used in a second run in the same conditions (adding the necessary amount of cellulose). In the second run, cellulose conversion was 45% and selectivity to sorbitol was 73%. This means that the catalyst preserves a relatively high activity and selectivity, which makes it reusable.

In a previous work dealing with the catalytic activity of SA, SASu and SAS carbons for cellulose hydrolysis [46], it was concluded that cellulose conversion and selectivity to glucose depend on the amount and type of surface OFG. In that study, the SAS carbon was found to lead to the highest cellulose conversion and glucose yield (61% and 52%, respectively). However, in the case of the Ru catalysts, the surface chemistry seems to have the opposite effect. As mentioned above, the presence of a high amount of OFG might lead to a less effective dispersion of the Ru particles, which could be associated with a lower activity. Moreover, some OFG, mainly those of acidic character (more abundant in Ru/SASu and Ru/SAS samples) may direct, likely assisted by the metal sites, the reaction to by-products through the conversion of glucose or sorbitol into other products. Furthermore, the presence of sulfonic groups, although in a low proportion, can contribute to acid catalyzed transformations that led to by-products.

Thus, in summary, the Ru/SA catalyst was found to be effective in enhancing the production of C6 polyols (sorbitol and mannitol), hindering the generation of by-products. It shows a good catalytic capability for the one pot hydrolytic hydrogenation of cellulose, similar or even higher than that shown by other catalysts reported in the literature, as some of those presented in the review work of Shrotri et al. [47] For comparison purposes, some examples of literature reported results obtained with carbon supported Ru catalysts (including reaction conditions) are summarized in Table 5 and commented on next.

Table 5. Summary of reported results in the hydrolytic hydrogenation of cellulose, using carbon supported Ru catalysts.

Entry	Cat. Name	Pret. [a]	S/Ru [b]	T (°C)	P (Bar)	T (h)	$X_{cellulose}$ (%)	$Y_{sorbitol}$ (%)	Ref.
1	Ru/BP2000	BM-2d	1620	190	50	24	86	37	[19]
2	Ru/AC(N)	-	324	190	9	3	51	17	[21]
3	Ru/AC(N)	MM-4d	324	190	9	3	89	58	[21]
4	Ru/CNT	H_3PO_4-MM	320	185	50	24	-	69	[22]
5	Ru/MC	-	556	190	50	1.5	35.4	20	[16]
6	Ru/CCD	BM-1d	25	180	40	10	32.7	-	[13]
7	Ru/CCD-SO$_3$H	BM-1d	25	180	40	10	100	63.8	[13]
8	Ru/AG-CNT	BM	556	205	50	5	100	60.4	[15]
9	Ru/SA	BM-7h	400	190	50	3	52	48	this work

[a] Cellulose pretreatment conditions; BM: ball milling, MM: Mix-milling, d: days, h: hours. [b] Substrate/Ru ratio (mg/mg). Ru was calculated taking into account the mass of catalyst used in the experiment and the wt. % Ru loading.

Kobayashi et al. [19] (entry 1) used a Ru catalyst supported on the BP2000 carbon black and high substrate (cellulose) to Ru ratio, and they achieved a very high cellulose conversion and an acceptable sorbitol yield but in a large reaction time (24 h). Komanoya et al. [21] (entry 2) used a Ru catalyst supported on an activated carbon and reported 51% cellulose conversion and 17% sorbitol yield using a relatively low hydrogen pressure (9 bar). However, when these authors applied mix milling (catalyst/cellulose) pretreatment (entry 3), both conversion and selectivity strongly increased (89% conversion and 58% sorbitol yield). This is a good result, but the long milling time (4 days) requires a lot of energy. Deng et al. [22] (entry 4) achieved a high sorbitol yield (69%) (cellulose conversion was not reported), using Ru supported on carbon nanotubes as catalysts. The reaction temperature was slightly lower than in the other examples, but cellulose was pretreated with a concentrated H_3PO_4 solution, and the reaction time was long (24 h). In the work of Zhang et al. [16] (entry 5), Ru was supported on a mesoporous carbon, but the obtained results were relatively poor; a significant improvement (95% cellulose conversion and 52% sorbitol yield) was achieved combining the Ru/MC catalysts with a zirconium phosphate (1.8 g per gram of cellulose). The recent work of Li et al. [13] that deals with Ru catalysts supported on carbonized cassava dregs (CCD), reported the need of the carbon sulfonation to achieve a high cellulose conversion and sorbitol yield (entries 6 and 7). Conditions of the process include, among others, harsh sulfuric acid treatment of the CCD material and high catalyst dosage. Finally, Rey-Raap et al. [15] (entry 8) reported very good sorbitol yield, using catalysts supported on a hybrid carbon material prepared with a glucose derived activated carbon and carbon nanotubes. The authors used ball-milled cellulose (milling conditions not indicated), 5 h reaction time, and a reaction temperature higher than in other reported works.

Regarding preparation conditions, with the exception of catalysts prepared by Kobayashi et al. [19] (entry 1) that use $Ru(NO)(NO_3)_3$; in all cases, $RuCl_3$ has been used as Ru precursor. Additionally, like in the present work, impregnation has been the general preparation procedure employed. It can be mentioned that, in the case of catalysts presented in entries 2 to 7, the catalysts reduction treatment temperature is higher than in the present work (300, 350 or 400 °C, vs. 250 °C), Rey-Raap et al. [15] also reduced the catalysts at 250 °C, and in the work of Kobayashi et al. [19], the catalysts were reduced at 180 °C, but suspended in water and under 4 MPa H_2, being then collected by filtration and dried, which adds some steps to the preparation process.

Some of the studies reported in Table 5 include reusability tests. After reaction, Li et al. [13] separated the mixture of catalyst and unreacted cellulose from the solution, and after washing and drying, fresh cellulose was added to start a second run. They repeated the process for up to five recycles, and found about a 10% decrease in sorbitol yield [13]. A slight sorbitol yield decrease (3%) was also observed by Komanoya et al. [21] in three reusing experiment tests, and Deng et al. [22] also found a decrease of the sorbitol yield (10%), especially after the first reuse. This behaviour is similar to the one found for the Ru/SA catalysts of the present work, in which sorbitol yield in the second run was about 15% lower than in the first one. However, Rey-Raap et al. [15] report very good reusability of their Ru catalyst supported on a hybrid carbon material (sorbitol yield kept above 60%).

To summarize, some interesting conclusions can be extracted, comparing our results with those dealing with carbon-based catalysts previously published. For example, the sorbitol yield obtained in the present work is higher, and it has been obtained in a much shorter time than in the study reported by Kobayashi et al. [19] (entry 1). The high cellulose conversion and the much lower sorbitol yield obtained by these authors indicate that their process is not very selective. The good results obtained by Komanoya et al. [21] (entry 3) require, as already mentioned, four days' mix-milling and, comparing the results of the present work with those of Deng et al. [22] (entry 4), a higher sorbitol yield has been obtained avoiding acid cellulose pretreatment, in a shorter reaction time and with a higher S/Ru ratio. The good results of Rey-Raap et al. [15] (entry 8) are undoubtedly very interesting, but compared with the present work, they have a required higher temperature and reaction time.

This comparison allows concluding that the catalysts prepared in this work are competitive in relation to other catalysts developed for the hydrolytic hydrogenation of cellulose, pointing out that both the catalysts' preparation and the reaction conditions used are mild, and can be considered as environmentally friendly.

4. Conclusions

The combined hydrolytic hydrogenation of cellulose into sorbitol was successfully achieved using catalysts prepared by supporting Ru nanoparticles (1 wt% Ru) on mesoporous carbon materials (the commercial activated carbon SA, and SASu and SAS carbons, both obtained by SA functionalization). The sulfonation treatment creates a moderated amount of oxygen functional groups on the carbon surface, and only a slight modification of the porous structure of the SA carbon. On the other hand, the oxidation treatment leads to an important development of surface functional groups, and a significant decrease in the adsorption capacity. Besides, the amount and type of the surface functional groups of the carbon materials becomes modified as a result of the Ru incorporation process. In spite of these modifications, it has been concluded that the mesopore volume remains high in the three catalysts. The Ru nanoparticles are small and well dispersed, and about 50% Ru is present in zero valent state even after air exposition of the catalysts.

The three catalysts exhibited high cellulose conversion and very good selectivity to sorbitol under relatively mild reaction conditions. The Ru/SA catalyst is the best performing one, with 52% cellulose conversion and a very high sorbitol selectivity (91%). The differences in surface chemistry seem to determine the observed differences in the catalytic behavior, and although a positive effect of a large amount of acidic OFG was foreseen, the results allow one to conclude that in the Ru containing catalysts, such groups can catalyze the formation of by-products, and are responsible for lower selectivity.

The findings of this study highlight the performance of the Ru/SA catalyst in the one pot hydrolytic hydrogenation of cellulose, which surpasses those of many other bifunctional catalysts. Moreover, it can be pointed out that both the catalyst preparation and the reaction conditions used in this work can be regarded as economically convenient and environmentally friendly.

Supplementary Materials: The following are available online at http://www.mdpi.com/1996-1073/13/17/4394/s1, Figure S1. Reactor scheme, Figure S2. Example of HPLC data obtained in the catalytic activity experiment carried out with the Ru/SA catalyst, Figure S3. N_2 adsorption isotherms at $-196\ ^{\circ}$C of the carbon supports and Ru catalysts, Figure S4. TPD profiles for supported Ru on carbon catalysts and carbons support, Figure S5. Deconvolution of the TPD profiles obtained for the supported Ru catalysts (SC: Strong acidic carboxyl groups, WC: Weak acidic carboxyl groups, CA: Carboxylic anhydride groups, CQ: Carbonyls or Quinones, PH: Phenol, LN: Lactones). Figure S6. TEM image of the SA activated carbon.

Author Contributions: Conceptualization, M.C.R.-M., M.Á.L.-R. and F.-Z.A.; methodology, M.C.R.-M., M.Á.L.-R. and F.-Z.A.; validation, M.C.R.-M., M.Á.L.-R. and F.-Z.A.; formal analysis, F.-Z.A.; investigation, F.-Z.A.; data curation, M.C.R.-M., M.Á.L.-R. and F.-Z.A.; writing—Original draft preparation, F.-Z.A.; writing—Review and editing, M.C.R.-M. and M.Á.L.-R.; supervision, M.C.R.-M. and M.Á.L.-R.; project administration, M.C.R.-M. and M.Á.L.-R.; funding acquisition, M.C.R.-M. and M.Á.L.-R. All authors have read and agreed to the published version of the manuscript.

Funding: This research was funded by Spanish Ministry of Science, Innovation and Universities and FEDER, project of reference RTI2018-095291-B-I00, GV/FEDER (PROMETEO/2018/076) and University of Alicante (VIGROB-136).

Acknowledgments: The authors thank funding to the Spanish Ministry of Science, Innovation and Universities and FEDER, project of reference RTI2018-095291-B-I00, GV/FEDER (PROMETEO/2018/076) and University of Alicante (VIGROB-136) for financial support. F.-Z. Azar thanks the AECID (research scholarship for development (2015/2016)) and University of Alicante (cooperation programs for development) for financial support.

References

1.	Satinder, K.B.; Saurabh, J.S.; Kannan, P. *Platform Chemical Biorefinery. Future Green Industry*, Brar, S.K., Sarma, S.J., Pakshirajan, K., Eds.; Elsevier: Amsterdam, The Netherlands, 2017; ISBN 9780128029800.

2.	Fang, Z.; Smith, R.L., Jr.; Li, H. (Eds.) *Production of Biofuels and Chemicals with Bifunctional Catalysts*; Springer: Berlin/Heidelberg, Germany, 2017.

3.	De Wild, P.; Reith, H.; Heeres, E. Biomass pyrolysis for chemicals. *Biofuels* **2011**, *2*, 185–208. [CrossRef]

4.	Sheldon, R.A. Green and sustainable manufacture of chemicals from biomass: State of the art. *Green Chem.* **2014**, *16*, 950–963. [CrossRef]

5.	Paone, E.; Tabanelli, T.; Mauriello, F. The rise of lignin biorefinery. *Curr. Opin. Green Sustain. Chem.* **2020**, *24*, 1–6. [CrossRef]

6.	Vilcocq, L.; Cabiac, A.; Especel, C.; Guillon, E.; Duprez, D. Transformation of sorbitol to biofuels by heterogeneous catalysis: Chemical and industrial considerations. *Oil Gas Sci. Technol.–Rev. d'IFP Energies Nouv.* **2013**, *68*, 841–860. [CrossRef]

7.	Awuchi, C.G. Sugar alcohols: Chemistry, production, health concerns and nutritional importance of mannitol, sorbitol, xylitol, and erythritol. *Int. J. Adv. Acad. Res. Sci. Technol. Eng.* **2017**, *3*, 31–66.

8. Brar, S.K.; Sarma, S.J.; Pakshijaran, K. Sorbitol production from biomass and its global market. In *Platform Chemical Biorefinery*; Elsevier: Amsterdam, The Netherlands, 2016; pp. 217–227.

9. Van de Vyver, S.; Geboers, J.; Jacobs, P.A.; Sels, B.F. Recent advances in the catalytic conversion of cellulose. *ChemCatChem* **2011**, *3*, 82–94. [CrossRef]

10. Besson, M.; Gallezot, P.; Pinel, C. Conversion of biomass into chemicals over metal catalysts. *Chem. Rev.* **2014**, *114*, 1827–1870. [CrossRef]

11. Rinaldi, R.; Schüth, F. Design of solid catalysts for the conversion of biomass. *Energy Environ. Sci.* **2009**, *2*, 610–626. [CrossRef]

12. Huang, Y.-B.; Fu, Y. Hydrolysis of cellulose to glucose by solid acid catalysts. *Green Chem.* **2013**, *15*, 1095–1111. [CrossRef]

13. Li, Z.; Liu, Y.; Liu, C.; Wu, S.; Wei, W. Direct conversion of cellulose into sorbitol catalyzed by a bifunctional catalyst. *Bioresour. Technol.* **2019**, *274*, 190–197. [CrossRef]

14. Gromov, N.V.; Medvedeva, T.B.; Taran, O.P.; Timofeeva, M.N.; Said-Aizpuru, O.; Panchenko, V.N.; Gerasimov, E.Y.; Kozhevnikov, I.V.; Parmon, V.N. The main factors affecting the catalytic properties of Ru/Cs-HPA systems in one-pot hydrolysis-hydrogenation of cellulose to sorbitol. *Appl. Catal. A Gen.* **2020**, *595*, 117489. [CrossRef]

15. Rey-Raap, N.; Ribeiro, L.S.; Órfão, J.J.d.M.; Figueiredo, J.L.; Pereira, M.F.R. Catalytic conversion of cellulose to sorbitol over Ru supported on biomass-derived carbon-based materials. *Appl. Catal. B Environ.* **2019**, *256*, 117826. [CrossRef]

16. Zhang, G.; Chen, T.; Zhang, Y.; Liu, T.; Wang, G. Effective Conversion of Cellulose to Sorbitol Catalyzed by Mesoporous Carbon Supported Ruthenium Combined with Zirconium Phosphate. *Catal. Lett.* **2020**, *150*, 2294–2303. [CrossRef]

17. Arunajatesan, V.; Chen, B.; Möbus, K.; Ostgard, D.J.; Tacke, T.; Wolf, D. Carbon-Supported Catalysts for the Chemical Industry. In *Carbon Materials for Catalysis*; Serp, P., Figueiredo, J.L., Eds.; John Wiley & Sons Inc.: Hoboken, NJ, USA, 2009; pp. 535–572, ISBN 9780470178850.

18. Besson, M.; Gallezot, P.; Perrard, A.; Pinel, C. Active carbons as catalysts for liquid phase reactions. *Catal. Today* **2005**, *102–103*, 160–165. [CrossRef]

19. Kobayashi, H.; Ito, Y.; Komanoya, T.; Hosaka, Y.; Dhepe, P.L.; Kasai, K.; Hara, K.; Fukuoka, A. Synthesis of sugar alcohols by hydrolytic hydrogenation of cellulose over supported metal catalysts. *Green Chem.* **2011**, *13*, 326–333. [CrossRef]

20. Manaenkov, O.V.; Kislitsa, O.V.; Matveeva, V.G.; Sulman, E.M.; Sulman, M.G.; Bronstein, L.M. Cellulose Conversion Into Hexitols and Glycols in Water: Recent Advances in Catalyst Development. *Front. Chem.* **2019**, *7*, 834. [CrossRef]

21. Komanoya, T.; Kobayashi, H.; Hara, K.; Chun, W.J.; Fukuoka, A. Kinetic study of catalytic conversion of cellulose to sugar alcohols under low-pressure hydrogen. *ChemCatChem* **2014**, *6*, 230–236. [CrossRef]

22. Deng, W.; Tan, X.; Fang, W.; Zhang, Q.; Wang, Y. Conversion of cellulose into sorbitol over carbon nanotube-supported ruthenium catalyst. *Catal. Lett.* **2009**, *133*, 167–174. [CrossRef]

23. Ribeiro, L.S.; Delgado, J.J.; Órfão, J.J.M.; Pereira, M.F.R. Carbon supported Ru-Ni bimetallic catalysts for the enhanced one-pot conversion of cellulose to sorbitol. *Appl. Catal. B Environ.* **2017**, *217*, 265–274. [CrossRef]

24. Chung, P.-W.; Charmot, A.; Click, T.; Lin, Y.; Bae, Y.; Chu, J.-W.; Katz, A. Importance of Internal Porosity for Glucan Adsorption in Mesoporous Carbon Materials. *Langmuir* **2015**, *31*, 7288–7295. [CrossRef]

25. Rodríguez-Reinoso, F.; Linares-Solano, Á. Microporous structure of activated carbon as revealed by adsorption methods. In *Chemistry and Physics of Carbon*; Marcel Dekker Inc.: New York, NY, USA, 1989; Volume 21, pp. 1–146.

26. Rouquerol, F.; Rouquerol, J.; Sing, K. *Adsorption by Powders & Porous Solids-Principles, Methodology and Applications*; Academic Press: London, UK, 1999.

27. Thommes, M.; Kaneko, K.; Neimark, A.V.; Olivier, J.P.; Rodríguez-Reinoso, F.; Rouquerol, J.; Sing, K.S.W. Physisorption of gases, with special reference to the evaluation of surface area and pore size distribution (IUPAC Technical Report). *Pure Appl. Chem.* **2015**, *87*, 1051–1069. [CrossRef]

28. Rufete-Beneite, M.; Román-Martínez, M.C.; Linares-Solano, A. Insight into the immobilization of ionic liquids on porous carbons. *Carbon* **2014**, *77*, 947–957. [CrossRef]

29. Figueiredo, J.L.; Pereira, M.F.R.; Freitas, M.M.A.; Órfão, J.J.M. Modification of the surface chemistry of activated carbons. *Carbon* **1999**, *37*, 1379–1389. [CrossRef]

30. Zhou, J.-H.; Sui, Z.-J.; Zhu, J.; Li, P.; Chen, D.; Dai, Y.-C.; Yuan, W.-K. Characterization of surface oxygen complexes on carbon nanofibers by TPD, XPS and FT-IR. *Carbon* **2007**, *45*, 785–796. [CrossRef]

31. Silva, A.M.T.; Machado, B.F.; Figueiredo, J.L.; Faria, J.L. Controlling the surface chemistry of carbon xerogels using HNO_3-hydrothermal oxidation. *Carbon* **2009**, *47*, 1670–1679. [CrossRef]

32. Li, N.; Ma, X.; Zha, Q.; Kim, K.; Chen, Y.; Song, C. Maximizing the number of oxygen-containing functional groups on activated carbon by using ammonium persulfate and improving the temperature-programmed desorption characterization of carbon surface chemistry. *Carbon* **2011**, *49*, 5002–5013. [CrossRef]

33. Adsuar-García, M.D.; Flores-Lasluisa, J.X.; Azar, F.Z.; Román-Martínez, M.C. Carbon-black-supported Ru catalysts for the valorization of cellulose through hydrolytic hydrogenation. *Catalysts* **2018**, *8*, 572. [CrossRef]

34. Machado, B.F.; Oubenali, M.; Rosa Axet, M.; Trang Nguyen, T.; Tunckol, M.; Girleanu, M.; Ersen, O.; Gerber, I.C.; Serp, P. Understanding the surface chemistry of carbon nanotubes: Toward a rational design of Ru nanocatalysts. *J. Catal.* **2014**, *309*, 185–198. [CrossRef]

35. Szymański, G.S.; Karpiński, Z.; Biniak, S.; Świątkowski, A. The effect of the gradual thermal decomposition of surface oxygen species on the chemical and catalytic properties of oxidized activated carbon. *Carbon* **2002**, *40*, 2627–2639. [CrossRef]

36. Figueiredo, J.L.; Pereira, M.F.R.; Freitas, M.M.A.; Órfão, J.J.M. Characterization of active sites on carbon catalysts. *Ind. Eng. Chem. Res.* **2007**, *46*, 4110–4115. [CrossRef]

37. Folkesson, B. ESCA Studies on the charge distribution in some dinitrogen complexes of Rhenium, Iridium, Ruthenium, and Osmium. *Acta Chem. Scand.* **1973**, *27*, 287–302. [CrossRef]

38. Shen, J.Y.; Adnot, A.; Kaliaguine, S. An ESCA study of the interaction of oxygen with the surface of ruthenium. *Appl. Surf. Sci.* **1991**, *51*, 47–60. [CrossRef]

39. KÖtz, R. XPS Studies of Oxygen Evolution on Ru and RuO2 Anodes. *J. Electrochem. Soc.* **1983**, *130*, 825–829. [CrossRef]

40. Oh, Y.J.; Yoo, J.J.; Kim, Y.I.; Yoon, J.K.; Yoon, H.N.; Kim, J.-H.; Park, S. Bin Oxygen functional groups and electrochemical capacitive behavior of incompletely reduced graphene oxides as a thin-film electrode of supercapacitor. *Electrochim. Acta* **2014**, *116*, 118–128. [CrossRef]

41. Velo-Gala, I.; López-Peñalver, J.J.; Sánchez-Polo, M.; Rivera-Utrilla, J. Surface modifications of activated carbon by gamma irradiation. *Carbon* **2014**, *67*, 236–249. [CrossRef]

42. Lin, Z.; Cai, X.; Fu, Y.; Zhu, W.; Zhang, F. Cascade catalytic hydrogenation–cyclization of methyl levulinate to form g -valerolactone over Ru nanoparticles supported on a sulfonic acid-functionalized UiO-66 catalyst. *RSC Adv.* **2017**, *7*, 44082–44088. [CrossRef]

43. Li, H.-X.; Zhang, X.; Wang, Q.; Zhang, K.; Cao, Q.; Jin, L. Preparation of the recycled and regenerated mesocarbon microbeads-based solid acid and its catalytic behaviors for hydrolysis of cellulose. *Bioresour. Technol.* **2018**, *270*, 166–171. [CrossRef]

44. Peng, G.; Gramm, F.; Ludwig, C.; Vogel, F. Effect of carbon surface functional groups on the synthesis of Ru/C catalysts for supercritical water gasification. *Catal. Sci. Technol.* **2015**, *5*, 3658–3666. [CrossRef]

45. Román-Martínez, M.C.; Cazorla-Amorós, D.; Linares-Solano, A.; De Lecea, C.S.-M.; Yamashita, H.; Anpo, M. Metal-support interaction in Pt/C catalysts. Influence of the support surface chemistry and the metal precursor. *Carbon* **1995**, *33*, 3–13. [CrossRef]

46. Azar, F.-Z.; Lillo-Ródenas, M.A.; Román-Martínez, M.C. Cellulose hydrolysis catalysed by mesoporous activated carbons functionalized under mild conditions. *SN Appl. Sci.* **2019**, *1*, 1739. [CrossRef]

47. Shrotri, A.; Kobayashi, H.; Fukuoka, A. Cellulose Depolymerization over Heterogeneous Catalysts. *Acc. Chem. Res.* **2018**, *51*, 761–768. [CrossRef] [PubMed]

 *energies*

Article

Investigating the Influence of Reaction Conditions and the Properties of Ceria for the Valorisation of Glycerol

Paul J. Smith, Louise Smith, Nicholas F. Dummer *, Mark Douthwaite, David J. Willock, Mark Howard, David W. Knight, Stuart H. Taylor * and Graham J. Hutchings

Cardiff Catalysis Institute, School of Chemistry, Cardiff University, Main Building, Park Place, Cardiff CF10 3AT, UK; Paul_John_Smith88@hotmail.com (P.J.S.); SmithL50@cardiff.ac.uk (L.S.); DouthwaiteJM@cardiff.ac.uk (M.D.); WillockDJ@cardiff.ac.uk (D.J.W.); HowardM1@cardiff.ac.uk (M.H.); KnightDW@cardiff.ac.uk (D.W.K.); hutch@cardiff.ac.uk (G.J.H.)

* Correspondence: dummernf@cardiff.ac.uk (N.F.D.); taylorsh@cardiff.ac.uk (S.H.T.)

Received: 13 March 2019; Accepted: 5 April 2019; Published: 9 April 2019

Abstract: The reaction of aqueous glycerol over a series of ceria catalysts is investigated, to produce bio-renewable methanol. Product distributions were greatly influenced by the reaction temperature and catalyst contact time. Glycerol conversion of 21% was achieved for a 50 wt.% glycerol solution, over CeO_2 (8 m^2 g^{-1}) at 320 °C. The carbon mass balance was >99% and the main product was hydroxyacetone. In contrast, at 440 °C the conversion and carbon mass balance were >99.9% and 76% respectively. Acetaldehyde and methanol were the major products at this higher temperature, as both can be formed from hydroxyacetone. The space-time yield (STY) of methanol at 320 °C and 440 °C was 15.2 and 145 g$_{MeOH}$ kg$_{cat}$$^{-1}$ h^{-1} respectively. Fresh CeO_2 was prepared and calcined at different temperatures, the textural properties were determined and their influence on the product distribution at *iso*-conversion and constant bed surface area was investigated. No obvious differences to the glycerol conversion or product selectivity were noted. Hence, we conclude that the surface area of the CeO_2 does not appear to influence the reaction selectivity to methanol and other products formed from the conversion of glycerol.

Keywords: ceria; glycerol; methanol; biodiesel

---

1. Introduction

The increased availability of glycerol; a by-product of first generation biodiesel production, has provided researchers with an opportunity to identify new routes to important platform chemicals and fuels from glycerol [1,2]. Production of biodiesel produces impure glycerol at approximately one tenth the mass of biodiesel [3] and consumes methanol derived from fossil fuels [4–6]. For this reason, it would be advantageous to develop a means of producing methanol directly from glycerol in a sustainable and economic manner.

To date, vapour phase reactions of dilute glycerol have been dominated by the solid acid catalysed production of acrolein [7–10]. The double dehydration of glycerol can produce acrolein in high yields and optimised catalysts can maintain this performance for >100 h [11]. Acrolein is a valuable intermediate for the production of acrylic acid [12], which is used in the manufacture of many plastics. Recently, Ueda and co-workers prepared an acid catalyst capable of producing acrylic acid from glycerol in one step [13].

Alternatively, hydroxyacetone [14,15], carbonates [16,17] and other polyols [17,18] can be produced from dilute glycerol, typically at reaction temperatures below 300 °C. The hydrogenolysis of glycerol to 1,2- and 1,3-propanediol in modest yields over CuO/ZnO [19], Rh based catalysts [20],

Raney Ni [21,22], and Cu-based catalysts [23] under high hydrogen partial pressures have also been reported. Copper-based catalysts have been successful in achieving high hydroxyacetone yields [14,15,24]. Nimlos and co-workers [25] proposed several reaction pathways for glycerol dehydration under both neutral and protonated forms. When protonated, the barrier to acrolein formation is substantially reduced and hydroxyacetone is formed in large quantities. They concluded that the barrier to glycerol dehydration under neutral conditions is high, and can therefore only occur at relatively high temperatures. Antal et al. [26] reported formation of the decomposition products acetaldehyde and acrolein, amongst other gaseous products, formed from glycerol in supercritical water at 500 °C. Further experiments were conducted to understand the formation of the liquid products with an acid catalyst and it was concluded that dehydration via a carbonium ion was enhanced by the presence of acids. In contrast, the acid catalyst did not enhance the formation of acetaldehyde via radical based homolytic C-C cleavage, which occurs readily at high temperatures [26].

The formation of products such as 1,2-propandiol, methanol and hydroxyacetone from dilute glycerol feeds over basic catalysts has been reported by Chai et al. [27]. They concluded that over CeO_2 and MgO, a high proportion of the product mixture was not identifiable and only small quantities of acrolein were observed. Furthermore, carbonaceous deposits were found to be 50 mg g_{cat}^{-1} over MgO at 315 °C, with 36.2 wt.% glycerol in water as the reaction solution. Velasquez et al. [24] reported the formation of methanol in very low yields (<1%) over a La_2CuO_4 catalyst. The yield of the main product, hydroxyacetone was found to be very sensitive to the oxidation state of the Cu contained in the mixed metal oxide. Over La_2O_3 the yield of hydroxyacetone was low and the carbon mass balance was less than 50%.

We recently developed a novel method to utilise both crude and refined glycerol to produce a crude methanol mixture [28]. This reaction of glycerol with water over very simple basic or redox oxide catalysts produced methanol, and other useful chemicals, in a one-step low pressure process without the addition of hydrogen gas. We proposed that methanol formed via a radical mechanism from some of the reaction intermediates; hydroxyacetone and ethylene glycol [28]. This paper details efforts to further investigate the process conditions and study the influence of the surface area of CeO_2 as redox catalysts, on their performance for glycerol conversion. Furthermore, the complex product mixture obtained over ceria and a total carbon content of a typical reaction is described in an attempt to close the carbon balance. The results presented here should then form the basis for future development work with ceria catalysts in order to achieve greater product yields to methanol, for example, through catalyst design.

2. Materials and Methods

2.1. Materials

Glycerol (≥99.5%), cerium(IV) hydroxide, cerium(III) nitrate hexahydrate (99.9% trace metals basis) were procured from Sigma-Aldrich (now Merck, Darmstadt, Germany). Ceria (CeO_2) (99.9% trace metal basis) from Acros Organics (now VWR, Radnor, PA, USA) and argon gas was purchased from BOC (Guildford, UK). These materials were used with no further treatments. Deionised (DI) water was provided in-house. Silicon carbide (SiC, 98%, Alfa Aesar, Ward Hill, MA, USA) of 40–50 mesh size was washed (DI water) and dried prior to use.

2.2. Catalyst Preparation

The precipitated ceria was prepared in the following way; a solution of cerium(III) nitrate hexahydrate was made (50 mL, deionised) and added to pre-heated deionized water under vigorous stirring (total 300 mL at 80 °C). The pH of the solution was monitored and ammonium hydroxide solution (1 M) was added dropwise until the pH reached 9. At this point, the slurry was immediately filtered and washed with warm deionised water (500 mL) and subsequently ethanol (200 mL).

The recovered solid was dried (120 °C, 16 h) and calcined under static air at 400, 500, 600 or 700 °C for 5 h, after a temperature ramp of 10 °C min^{-1} from ambient.

2.3. Catalyst Testing

Catalytic reactions were carries out using a gas-phase micro-reactor operating under continuous flow. Aqueous solutions containing glycerol (50 wt. %) were fed using an HPLC pump at flow rates of 0.016–0.048 mL min^{-1} into a preheater and vaporised (305 °C). The glycerol vapour passed through the reactor using a carrier gas; argon (15–45 mL min^{-1}). The reactor pipes were heated to prevent condensation of feedstock or reaction products. Catalysts were used with a uniform particle size (250–425 μm) and formed through pelleting, crushing and finally sieved. Typically, the catalyst samples (0.5 g) were diluted with silicon carbide to a uniform volume (2 mL) and placed into a stainless steel tube with an 8 mm inner diameter supported by quartz wool, above and below the bed. These conditions resulted in mass velocities and space velocities between 675–5400 L h$^{-1}_{Ar}$ kg$^{-1}_{cat.}$ and 1580–10,800 L h$^{-1}_{Ar}$ L$^{-1}_{cat.}$ respectively. A thermocouple was placed in the catalyst bed and used to control reaction temperature, typically between 320–480 °C. Reaction products (liquids) were collected using a stainless steel trap (held at ca. 0 °C). Gaseous products were collected in a gas bag that was attached at the exit line of the liquid trap.

Analysis of Liquid reaction products were performed offline using a CP 3800 gas chromatograph (GC1, Varian now Agilent Technologies, Santa Clara, CA, US; capillary column; ZB-Wax plus, 30 m × 0.53 mm × 1 μm). An external standard (cyclohexanol) was used. Gaseous, carbon based reaction products were analysed offline using a Varian 450-GC gas chromatograph (GC2; capillary column; CP-Sil5CB, 50 m × 0.32 mm × 5 μm). Non-carbon based gases; H_2 and O_2 were analysed using a Varian CP3380 gas chromatograph (GC3; Porapak Q column). Product selectivities (carbon mol. %) were calculated from the moles of carbon in a product recovered divided by the total moles of carbon in all products. Product yield (mol. %) was calculated from the moles of product recovered divided by the total number of moles of glycerol injected. Product list and retention times based on GC analysis is illustrated in Table S1. Additional qualitative analysis of the post reaction liquid sample was achieved with liquid chromatography-mass spectrometry (LCMS). Analysis was performed with a Bruker Amazon SL ion trap mass spectrometer, operated in a positive electrospray ion mode and paired to an Ultimate HPLC system (Thermo Fisher Scientific, Waltham, MA, USA). The HPLC (C-18 column and maintained at 40 °C) analysis used a gradient elution, consisting of 0.1% formic acid in H_2O (A) and 0.1% formic acid in acetonitrile. The gradient elution was performed as reported in Table 1 on 10 μL samples of the reaction mixture.

Table 1. The makeup of the mobile phase for the gradient elution.

Time (min)	A [1] (%)	B [2] (%)
0.0	98	2
1.0	98	2
15.0	2	98
17.0	2	98
18.0	98	2
20.0	98	2

[1] A = 0.1% formic acid in H_2O and [2] B = 0.1% formic acid in acetonitrile.

2.4. Calculations

The glycerol conversion (C_{GLY}) was calculated according to Equation (1) and based on the molar difference between carbon from glycerol fed into the reactor, g_{mi}, and that detected at the outlet, g_{mo}:

$$C_{GLY} \ (\%) = \left(\frac{g_{mi} - g_{mo}}{g_{mi}} \right) \times 100 \tag{1}$$

The product selectivity ($S_p(x)$, carbon mol. %) for any product, x, was calculated from the moles of carbon recovered in x, x_{Cm} divided by the sum of moles of carbon in each product, y_{Cm} (Equation (2)):

$$S_p(x)(\%) = \left(\frac{x_{Cm}}{\Sigma_y\, y_{Cm}} \right) \times 100 \tag{2}$$

The carbon balance can be obtained by comparing the moles of carbon accounted for in unreacted glycerol and in the identified products to the moles of carbon in glycerol entering the reactor:

$$B_C(\%) = \left(\frac{g_{mo} + \Sigma_x\, x_{Cm}}{g_{mi}} \right) \times 100 \tag{3}$$

Functional group yield (Y, carbon mol. %) data were calculated by the sum of products containing that functional group as a function of their selectivities (S_G), multiplied by conversion C_{GLY}, multiplied by the carbon balance B_C, omitting coke (Equation (3)).

$$Y\,(\%) = \left(\frac{(\Sigma\, S_G) \times C_{GLY}}{100} \right) \times B_C\,(\%) \tag{4}$$

The overall carbon balance B_{Ctot} was calculated (Equation (5)) by dividing the sum of the carbon moles of products x_{Cm}, coke x_{Ccoke} estimated from post reaction characterisation and unreacted glycerol g_{mo} by the carbon moles of glycerol injected into the reactor g_{mi}:

$$B_{Ctot} = \left(\frac{\Sigma_x\, x_{Cm} + x_{Ccoke} + g_{mo}}{g_{mi}} \right) \times 100 \tag{5}$$

The hydrogen balance B_H was calculated (Equation (6)) by dividing the sum of the hydrogen moles of products x_H, hydrogen gas (GC3) x_{Hgas} and moles of hydrogen in unreacted glycerol g_{Hmo} by the moles of hydrogen in glycerol injected into the reactor g_{Hmi}.

$$B_H = \left(\frac{x_{Hp} + x_{Hgas} + g_{Hmo}}{g_{Hmi}} \right) \times 100 \tag{6}$$

The oxygen balance B_O was calculated (Equation (7)) by dividing the sum of the oxygen moles of products x_O, oxygen gas (GC3) x_{Ogas} and moles of oxygen in unreacted glycerol g_{Omo} by the moles of oxygen in glycerol injected into the reactor g_{Omi}:

$$B_O = \left(\frac{x_O + x_{Ogas} + g_{Omo}}{g_{Omi}} \right) \times 100 \tag{7}$$

Carbon deposition, referred to as coke on the catalyst was calculated dividing the mass loss as analysed by TGA of the used catalyst m_{LOST}, by the carbon moles of glycerol feed over the catalyst g_{mi} (Equation (8)):

$$Coke\,(\%) = \left(\frac{m_{LOST}}{g_{mi}} \right) \times 100 \tag{8}$$

The methanol space-time-yield STY_{MeOH}, was calculated (Equation (9)) from the mass of methanol m_{MeOH}, produced per h (reaction time Rt), per mass of catalyst (m_{cat}, kg):

$$STY_{MEOH} = \left(\frac{m_{MEOH}\,(g)}{Rt\,(h) \times m_{cat}\,(kg)} \right) \tag{9}$$

2.5. Catalyst Characterisation

Powder X-ray diffraction (XRD) analysis of the catalysts was carried out on a PANalytical X'pert Pro powder diffractometer (Malvern Panalytical, Malvern, UK) using a Cu source operated at 40 KeV and 40 mA with a Ge (111) monochromator to select $K_{\alpha 1}$ X-rays. Patterns were analysed from measurements taken over the 2θ angular range 10–80° (step size of 0.016°).

Thermal gravimetric analysis (TGA) and differential thermal analysis (DTA) were carried out using a Labsys 1600 instrument (Setaram, Caluire, France). Alumina crucibles containing the samples (20–50 mg) were loaded into the instrument and heated to 900 °C (5 °C/min) under synthetic air (50 mL min^{-1}). To exclude buoyancy effects, for specified TGA runs a blank experiment was subtracted from the relevant data.

Surface area analysis was carried out using the Brunauer Emmett Teller (BET) method with a QUADRASORBevo™ surface area and pore size analyser (Quantachrome a brand of Anton Parr, Boynton Beach, FL, USA). A forty point (20 adsorption and 20 desorption points) analysis was carried out using an adsorbate gas (N_2 at -196 °C). Samples (300 mg) were degassed under vacuum for 3 h at 110 °C prior to analysis.

Raman spectroscopy was performed using an inVia microscope (Renishaw, Gloucestershire, UK) operated at a wavelength of 514 nm. 10 acquisitions were performed per sample with an exposure time of 10 s; the laser was employed at 1% power.

3. Results and Discussion

3.1. Influence of Contact Time and Reaction Temperature

To ensure consistency across batches, commercially available CeO_2, obtained from Acros Organics, (8 m^2 g^{-1}) was used as a catalyst to examine the influence of contact time and reaction temperature on the conversion of glycerol. Figure 1 illustrates the effect of the contact time on the product yield per gram of the catalyst, for reactions at 320 °C with a 50 wt.% aqueous glycerol feed, over 0.5, 1.5 and 4 g of CeO_2. Increasing the catalyst mass, increased the bed volume accordingly, which was determined to be 0.25, 0.65 and 1.7 mL for the reactions with 0.5, 1.5 and 4 g of CeO_2 respectively. This consequentially effected the GHSV, which ranged from 10,800 to 1588 h^{-1}. The glycerol conversion increased from 5, to 21, to 56% with the increasing catalyst bed volume. The carbon mass balance remained high (>99%) over the 0.5 g and 1.5 g catalyst samples. Over 4 g of catalyst, the mass balance was 92%. We consider the high carbon mass balances observed are due to the low relative glycerol conversion and the nature of the products formed, i.e., a low yield of aldehydic molecules.

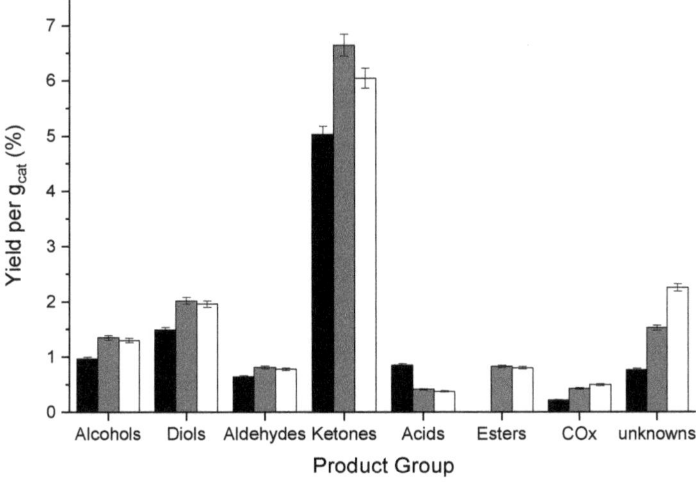

Figure 1. Influence of contact time on the yield of product groups per gram of catalyst over 0.5 g (black bar, conversion 5%), 1.5 g (grey bar, conversion 21%) and 4 g (white bar, conversion 56%) of ceria at 320 °C. Respective GHSV; 10800 h^{-1}, 4154 h^{-1} and 1588 h^{-1}.

Comparing the product selectivities from the experiments collected in Figure 1 is challenging, given that the glycerol conversions observed are notably different. These reactions do however, offer a means to understand the broad influence of catalyst contact time. The full product distributions for these experiments are complex and the full range of products are displayed in the supplementary information (Table S2). Over 1.5 g of ceria, the main products observed were hydroxyacetone, 1,2-propanediol and methanol with carbon mole selectivities of 45.4%, 6.0% and 6.0% respectively. Here we classify hydroxyacetone as a ketone despite its dual functionality, according to the most oxidised group taking precedent. Previously [28,29], we reported that a number of the products, including; methanol, acetaldehyde and 2,3-butanedione, originate from secondary reactions, which we proposed were formed from hydroxyacetone. As such, it would be rational to consider that as the contact time increases, the abundance of these products should also increase. However, in general the product yields per gram of catalyst remained similar.

Over 1.5 g and 4 g of ceria, the yield per gram of hydroxyacetone was slightly reduced and an increase in the yield of an ester was also observed, along with a modest decrease in the yield of carboxylic acids. This result suggests that the catalyst surface, may promote the conversion of acids to esters. Interestingly, when the product yields are normalised to account for changes in the catalyst mass, the observed loss of acid products does not appear to account for the significant increase in esters and other unidentified products that could now be present (Figure 1). However, we consider that the modest reduction of hydroxyacetone selectivity in combination with the loss of acids could account for the generation of esters. Chen et al. [30] reported the possibility of the esterification of a primary alcohol with a carboxylic acid, which suggests that hydroxyacetone reacted with acetic acid for example. Consequently, the increase in the yield of 2-oxopropyl acetate (Scheme 1), can be related to the increased contact time over high mass catalyst beds, where products can react further.

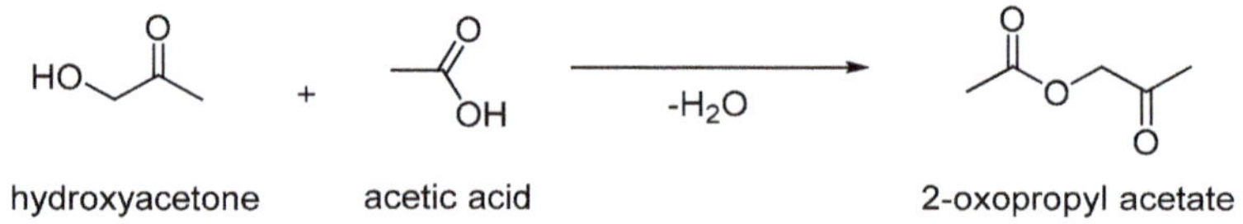

HO   +   OH   -H$_2$O   →   2-oxopropyl acetate

hydroxyacetone   acetic acid   2-oxopropyl acetate

Scheme 1. Condensation reaction of hydroxyacetone and acetic acid to form 2-oxopropyl acetate. Similar reactions could account for the increase in unidentifiable products at high contact times over CeO_2.

The methanol STYs calculated from the experiments conducted over 0.5 and 1.5 g of ceria is comparable; 15.8 and 15.2 g_{MeOH} kg_{cat}^{-1} h^{-1} respectively. However, over 4 g of ceria this reduces to 11.6 g_{MeOH} kg_{cat}^{-1} h^{-1}, which we attribute to the increased contact time and an increased potential for product re-adsorption and subsequent reaction, demonstrated by the increased quantity of unknowns produced. In contrast, the space-time-yield of hydroxyacetone was comparable over 1.5 and 4 g of ceria, and calculated to be 84.2 and 85.5 $g_{hydroxyacetone}$ kg^{-1} and h^{-1} respectively, compared with 63.3 $g_{hydroxyacetone}$ kg^{-1} h^{-1} over 0.5 g of ceria.

A reaction temperature of 320 °C, does not appear to be high enough to significantly promote secondary reactions involving hydroxyl containing intermediates, such as hydroxyacetone, ethylene glycol or 1,2-propanediol, which can undergo sequential reactions that lead to the formation of methanol. As we noted previously [28], the STY of methanol is highly dependent on the reaction temperature, which we previously postulated to be due to the conversion of intermediate products such as hydroxyacetone. Table 2 contains the product yields, and Figure 2 the product selectivities collated into major functional groups, as a function of the reaction temperature. The individual product selectivities from these reactions are given in Table S3. At lower reaction temperatures; below 400 °C, the formation of hydroxyacetone, 1,2-propanediol and ethylene glycol predominate. At reaction temperatures above 400 °C, the yield of ketones decreases from 18.3 to 12.5%, and the production of alcohols and aldehydes increases from 16.6 to 21.6% and 18.8 to 30.7% respectively. The yield of CO$_x$

also increases to 11.6% and the carbon mass balance decreases to 77% as the reaction temperature is increased.

Very little hydroxyacetone was observed for the reactions conducted at temperatures above 400 °C. However, increases in the quantity of acetone and 2,3-butanedione were observed as the reaction temperature was raised. These products are also considered to originate from hydroxyacetone, which correlates with the substantial increase in the quantity of both acetaldehyde and methanol [28]. Methanol accounts for the vast majority of the total alcohol products and is considered to be relatively stable at the higher reaction temperatures used (Table S4). The quantity of diols; 1,2-propanediol, ethyl glycol and 1,3-propanediol, also reduced in a similar fashion to hydroxyacetone (<1% selectivity) over the catalyst at 440 °C. The STY of methanol increases when the reaction temperature is increased; from 15.8 to 59.8 to 100.2 to 145.7 g_{MeOH} kg_{cat}^{-1} h^{-1} at 320, 360, 400 and 440 °C respectively. The differences in the product distributions at lower and high reaction temperatures strongly suggest that many of the C_2+ oxygenates, predominantly formed at low temperatures, are likely to be reaction intermediates. More specifically, it appears that C_2 and C_3 molecules containing one or more hydroxyl groups can undergo both C-O and C-C dissociation at higher temperatures.

Table 2. Glycerol conversion and product distribution for reactions over CeO_2 at different temperatures.

RT [a]/°C	C_{GLY} [b]/%	Mol. Balance [c]/%			Yield/% [e]							
		B_C [d]	B_H	B_O	Alc.	Diols	Ald.	Ket.	Ac.	Est.	CO_x	Unk.
320	21	100	97	95	2.0	3.1	3.1	10.6	1.8	0.0	0.5	1.6
360	84	87	78	72	10.1	12.9	12.9	29.4	6.6	1.7	3.7	13.0
400	98	80	68	62	16.6	6.9	6.9	18.3	10.4	2.0	6.7	18.1
440	100	77	62	57	21.6	1.3	1.3	12.5	6.3	1.3	11.6	14.3

[a] Reaction temperature; [b] Glycerol conversion; [c] Carbon, hydrogen and oxygen mass balance (±3%) of products detected in GC1 and GC2; [d] Carbon balance, values in parenthesis include coke from TGA measurements; [e] yield of products by functional group detected in GC1 and GC2; Alc., alcohols; Ald., aldehydes; Ket., ketones; Ac., acids; Est., esters; Unk., unidentified products (Full product list in Table S2). Reaction conditions; CeO_2 (8 m^2 g^{-1}) 1.5 g, GHSV 4154 h^{-1}, 50 wt.% Gly/H_2O, Ar 45 mL min^{-1}.

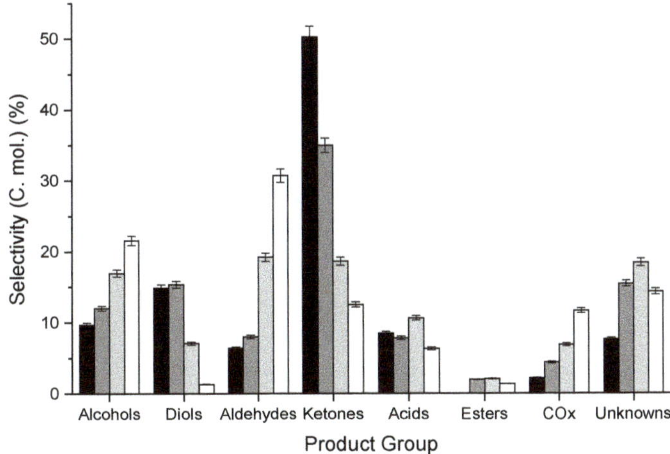

Figure 2. Influence of the reaction temperature on the product group selectivity at 320 °C (black bar), 360 °C (dark grey), 400 °C (light grey) and 440 °C (white bar). Reaction conditions; CeO_2 (8 m^2 g^{-1}) 1.5 g, GHSV 4154 h^{-1}, 50 wt.% Gly/H_2O, Ar 45 mL min^{-1}.

The decrease of the carbon mass balance from 100 to 77% as the reaction temperature increases, coincides with an increase in the quantity of aldehydes observed. The stability of aldehydes over ceria has been studied on polycrystalline materials [31] and thin films with a (111) surface [32,33].

These studies indicated that the ceria surface can dramatically influence the acetaldehyde reactivity. Idriss et al. [31] reported that acetaldehyde can undergo many reactions on the surface of ceria, which can result in the desorption of products such as CO_2, CH_4 and C_4 olefins at similar temperatures to those used in the present study (400 °C). Acetaldehyde has been reported to chemisorb weakly on oxidised ceria, however, it was found to bind strongly as a carbanion to reduced ceria [32]. This carbanion species will likely partake in sequential reactions to form other products or be retained on the surface as coke, and hence, lower the carbon mass balance. Surface-bound organic molecules or coke, only contribute to approximately 2% of the overall carbon mass balance, which was determined from TGA of the post-reaction samples.

3.2. Influence of Catalyst Calcination Temperature

Low surface area ceria has been shown to be an effective catalyst for the conversion of glycerol to methanol, but typically requires a high reaction temperature (>400 °C). However, at these temperatures the selectivity to methanol is compromised by an increased production of aldehydic molecules such as acetaldehyde (Table S3). Modifications to the ceria surface area may allow for a greater methanol selectivity at lower reaction temperatures. For this reason, the influence of ceria calcination temperature on product selectivity was investigated, and the influence of the surface area and crystallite size with respect to the catalytic activity and methanol STY was established. A cerium hydroxide catalyst precursor was prepared from $Ce(NO_3)_3 \cdot 9(H_2O)$ by precipitation with NH_4OH. The precursor was calcined at 400, 500, 600 and 700 °C and characterised by powder XRD, Raman spectroscopy and N_2 adsorption; to determine the crystallite size, defect density and BET surface area respectively. These materials were subsequently tested at 340 °C with a 50 wt.% aqueous glycerol feed.

The powder XRD patterns of the CeO_2 materials calcined at different temperatures are displayed in Figure 3. All the reflections can be indexed to a cubic fluorite structure (JCPDS no. 34-0394) and are characteristic of an Fm-3m space group. The intensity of diffraction peaks increase and the peak widths decrease with increasing calcination temperature, and consequently the crystallite size increases (Table 3). The crystallite size was calculated with the Scherrer equation, however, we take the values as a guide due to the cubic nature of the CeO_2 particles within this polycrystalline material. Mamontov et al. also observed an increase in crystallite size according to an increasing calcination temperature, and determined that this change was permanent upon cooling [34]. Concurrent with this increase of crystallite size is an expected loss of surface area. The ceria calcined at 400 °C exhibited a surface area of 38 $m^2\ g^{-1}$ which decreases to 22 $m^2\ g^{-1}$ when calcined at 700 °C (Table 3). Additionally, the defect density generally decreases according to the increasing calcination temperature (Table 3). The defect density was calculated from Raman spectroscopy by considering the ratio of the peak areas at ca. 465 cm^{-1} (F_{2g}) and the mode at ca. 590 cm^{-1}, which indicates the presence of intrinsic oxygen vacancies giving rise to a Ce^{3+} defect mode [35,36]. The F_{2g} mode represents the symmetrical stretching of the oxygen atoms about Ce^{4+}. The ratio of the Raman lines can be used to estimate the defect density. The material calcined at 400 °C has a defect density of 0.58%, which increases to 0.83% when calcined at 500 °C. At higher calcination temperatures, this decreases significantly to 0.05% after calcination at 700 °C, and can be attributed to the increase in the crystallinity of the structure. The intensity of the F_{2g} mode initially decreases with increasing calcination temperature, however, when calcined at 700 °C the intensity significantly increases due to the restructured crystallites. The defect density has been strongly linked to redox catalytic activity of ceria through its oxygen storage capacity [35,37,38]. All the ceria materials in this study, typically possess low quantities of defects (Table 3) compared to CeO_2 materials doped with rare-earth metals such as Pr^{3+} or Sm^{3+} [39].

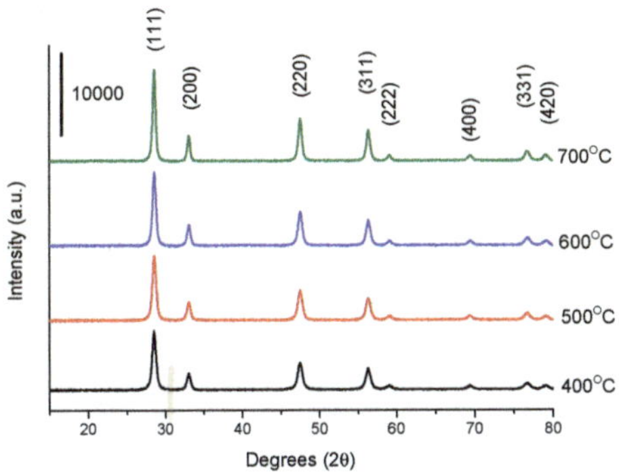

Figure 3. Powder XRD pattern of CeO_2 calcined at different temperatures (400–700 °C); reflections match the cubic fluorite structure of ceria (Fm-3m).

Table 3. Textural properties of CeO_2 catalysts calcined at different temperatures.

Calcination Temperature/°C	BET Surface Area [a]/m² g⁻¹	Crystallite Size [b]/nm	Defect Density [c]/%	Defect Density per Surface Area [c]/×10⁻³% m⁻²	FWHM [d] of F_{2g} Mode
400	38	14	0.58	1.5	17.96
500	34	15	0.83	2.4	18.03
600	27	16	0.11	0.4	16.61
700	22	19	0.05	0.3	13.35

[a] calculated from N_2 adsorption measurements at −196 °C (adsorption/desorption isotherm Figure S1); [b] calculated from the Scherrer equation using the (111) reflection in Figure 3; [c] Calculated from Raman spectroscopy in Figure 4 by I_D/I_{F2g} where I_D is the integrated area of the band at ca. 590 cm⁻¹ and I_{F2g} is the integrated area of the F_{2g} band at 460 cm⁻¹; [d] Full width at half maximum.

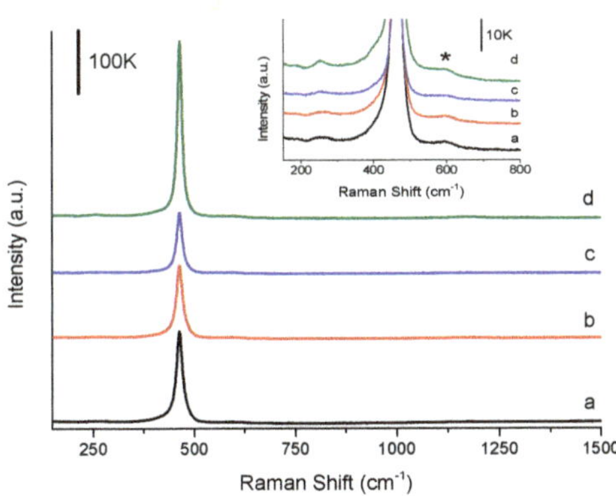

Figure 4. Raman spectra of the ceria samples calcined at different temperatures with F_{2g} mode at ca. 465 cm⁻¹; (a) 400, (b) 500, (c) 600 and (d) 700 °C. Insert highlights the relative intensity of the defect mode (*) at ca. 590 cm⁻¹.

The corresponding glycerol conversions and product yields for ceria catalysts calcined at varying temperature are displayed in Table 4. These reactions were carried out using differing catalyst masses to maintain a consistent catalyst bed surface area. Furthermore, the argon gas flows was adjusted such that the GHSV in each of the experiments was comparable; approximately 3450 h⁻¹ based on a uniform catalyst bed volume. The glycerol partial pressure changed from 47 to 29 Pa over the catalysts in Table 4, and represents ca. 0.05% of the total system pressure. The glycerol conversion was ca. 83% over each of the CeO_2 catalysts tested (Table 4). Therefore, if any changes in the product selectivity (Figure 5) are observed, this can be assigned to surface features of the different CeO_2 catalysts. We reported previously on the influence of the SiC diluent and we consider that the minor changes to the mass of SiC in this set of experiments will not significantly impact the product distribution [29]. The collective product yields produced in these experiments are listed in Table 4 and are comparable for each of the catalysts tested.

The carbon mole selectivity to hydroxyacetone over each of the catalysts calcined at 400–700 °C was 32.1, 33.0, 28.8 and 29.1% respectively, with a standard deviation of 1.8% (Table S5). The deviation of other major products such as acetaldehyde, methanol and ethylene glycol was also minimal, and calculated to be <1%. Therefore, we consider the reaction selectivities reported in Figure 5 and Table S5 to be comparable. The carbon mass balance across the samples tested was similar (*ca.* 82%), and for all catalysts the degree of fouling by carbon deposition was low (*ca.* 20 mg_{Coke} g^{-1}). This result implies that the surface area differences, crystallite size and defect density do not appear to direct the product selectivities at *iso*-conversion. This observation suggests that the surface crystallite facets may have more influence on product selectivity and work is underway to investigate this.

Carbon deposition on the catalyst during the reactions was determined to be insignificant (<3%), even when reactions were conducted at high temperatures. Total organic carbon analysis was performed on the post-reaction effluent and was determined to be 95% over the CeO_2 catalyst calcined at 600 °C. As such, we conclude that the remaining 5% of the missing carbon is attributed to reactor fouling. The carbon mass balance based on liquid and gas phase products, as quantified by GC1 and 2, along with any catalytic carbon deposition, was 85%, with 79% comprised of liquid phase products (Table S6). The inference is that chemical species are present which cannot be analysed by GC totalling ca. 10% of the carbon moles injected into the reactor. Consequently, LC-MS analysis was utilised to qualitatively monitor the presence of species with high molecular weights. Fragments were analysed between 100 and 1000 m/z. The reaction performed at 340 °C, over 0.7 g of ceria calcined at 600 °C, was analysed by LC-MS (Figure S2). A blank sample (D.I. water) was also analysed to exclude any column bleed, which may be observed at the end of the trace. The corresponding chromatogram was very complex, with numerous peaks detected, making product identification very challenging. However, whilst specific products cannot be identified, the use of LC-MS confirms the presence of high molecular weight fragments, which may form through base-catalysed condensation reactions, indicating the presence of undesirable side products. Therefore, we attribute the difference between the TOC observed post reaction (95%) and the corresponding carbon mass balance (85%) to a proportion of heavy products produced during the reaction, which are not visible by GC.

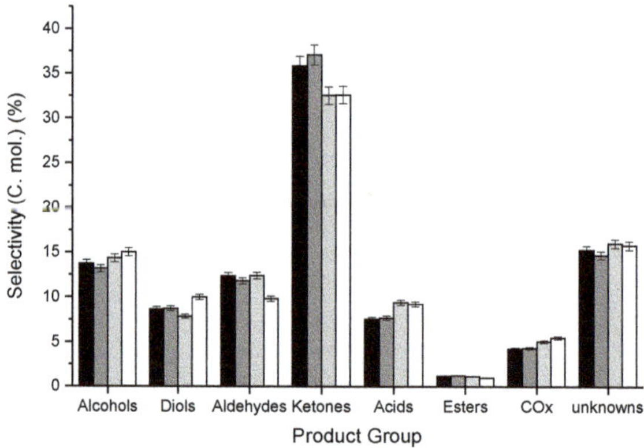

Figure 5. The product group selectivity over CeO_2 calcined at different temperatures where the catalyst mass and Ar flow rate has been adjusted to maintain a consistent catalyst bed surface area; 400 °C (black bar), 500 °C (dark grey), 600 °C (light grey) and 700 °C (white bar). Reaction conditions; 340 °C, 50 wt.% Gly/H_2O, GHSV ca. 3450 h^{-1}.

Table 4. Glycerol conversion and product yield distribution for reactions over CeO_2 calcined at different temperatures (CT) at a uniform catalyst bed surface area.

CT/°C	Catalyst Mass/g	Ar Flow Rate/mL min^{-1}	C_{GLY} [a]/%	Mol. Balance/% [b]			Yield [d]/%								STY_{MeOH} [e]/g^1 h^{-1} kg^{-1}	Carbon Deposition/mg$_{Coke}$ g^{-1}
				B_C [c]	B_H	B_O	Alc.	Diols	Ald.	Ket.	Ac.	CO_x	Est.	Unk.		
400	0.4998	15	83	82 (83)	75	67	7.9	6.0	8.5	24.7	5.2	2.9	0.7	11.3	84.3	17
500	0.5497	15	83	80 (81)	72	66	7.4	5.9	7.9	24.9	5.2	2.9	0.7	10.7	80.0	21
600	0.6995	20	80	84 (86)	76	70	8.1	5.3	8.4	25.2	6.4	3.4	0.6	11.7	59.3	19
700	0.8508	25	84	81 (84)	74	67	7.7	6.9	6.8	22.4	6.3	3.8	0.5	11.5	51.1	22

[a] Glycerol conversion; [b] Carbon, hydrogen and oxygen mass balance ($\pm$3%) of products detected in GC1 and GC2; [c] Carbon balance, values in parenthesis include coke from TGA measurements; [d] yield of products detected in GC1 and GC2; Alc., alcohols; Ald., aldehydes; Ket., ketones; Ac., acids; Est., esters; Unk., unidentified products (Full product list in Table S5); [e] methanol space time yield. Reaction conditions; 340 °C, 50 wt.% Gly/H_2O, GHSV ca. 3450 h^{-1}.

4. Conclusions

The conversion of glycerol over ceria catalysts is a complex process, consisting of numerous reaction routes which can result in a wide variety of products. Both the temperature and catalyst contact time heavily influence the product distribution, with higher temperatures required to achieve higher space-time-yields of methanol. Increasing reaction temperatures resulted in a reduction in hydroxyacetone, potentially via its participation in sequential catalytic reactions; leading to increased quantities of acetaldehyde, CO_x, coke and other undetectable products. A subsequent investigation was conducted in order to establish the relationship between some physicochemical properties of CeO_2 and product distribution. In order to test these materials at comparable GHSVs, the quantity of catalyst used in these experiments was normalised to account for their variations in surface area. There appears to be no clear relationship between the quantity of CeO_2 defect sites and the reactivity of glycerol and its intermediates. As such, we propose that the reactivity of glycerol and its intermediates in this reaction are predominantly driven by the morphology of the catalyst, which will provide a foundation for future work in this area.

Supplementary Materials: The following are available online at http://www.mdpi.com/1996-1073/12/7/1359/s1, Figure S1: Nitrogen adsorption and desorption isotherms for ceria calcined at 400, 500, 600 and 700 °C, Figure S2: LC-MS chromatogram corresponding to the post reaction solution of a reaction run over CeO_2 for 6 h. Table S1: Liquid and gas product list, Table S2: Influence of contact time on dilute glycerol reaction, Table S3: Influence of reaction temperature on dilute glycerol reaction, Table S4. Methanol stability over commercial CeO_2 at 400°C, Table S5: Influence of CeO_2 calcination temperature on dilute glycerol reaction with GHSV ca. 3450 h^{-1}, Table S6: The total carbon content (%) observed in a reaction over CeO_2 for 6 h at GHSV ca. 3600 h^{-1}.

Author Contributions: Conceptualization, N.F.D., S.H.T. and G.J.H.; Data curation, P.J.S. and L.S.; Investigation, D.W.K.; Methodology, N.F.D., M.H. and S.H.T.; Supervision, D.J.W., S.H.T. and G.J.H.; Writing—original draft, N.F.D. and S.H.T.; Writing—review & editing, M.D., D.J.W., D.W.K. and G.J.H.

Funding: This research was funded by EPSRC, grant number EP/P033695/1 and EP/L027240/1.

Acknowledgments: The authors would also like to thank Exeter Analytical UK Ltd. for the Total Organic Content analysis. We would also like to acknowledge Thomas Williams for his assistance with the operation of and processing of the LC-MS data. Information on the data underpinning the results presented here can found in the Cardiff University data catalogue at http://doi.org/10.17035/d.2019.0067888810.

References

1. Gu, Y.; Jerome, F. Glycerol as a sustainable solvent for green chemistry. *Green Chem.* **2010**, *12*, 1127–1138. [CrossRef]

2. Sun, D.; Yamada, Y.; Sato, S.; Ueda, W. Glycerol as a potential renewable raw material for acrylic acid production. *Green Chem.* **2017**, *19*, 3186–3213. [CrossRef]

3. Otera, J. Transesterification. *Chem. Rev.* **1993**, *93*, 1449–1470. [CrossRef]

4. Twigg, M.V.; Spencer, M.S. Deactivation of supported copper metal catalysts for hydrogenation reactions. *Appl. Catal. A* **2001**, *212*, 161–174. [CrossRef]

5. Wainwright, M.S. Catalytic processes for methanol synthesis—Established and future. *Stud. Surf. Sci. Catal.* **1988**, *36*, 95–108.

6. Thompson, J.C.; He, B.B. Characterization of crude glycerol from biodiesel production from multiple feedstocks. *Appl. Eng. Agric.* **2006**, *22*, 261. [CrossRef]

7. Katryniok, B.; Paul, S.; Capron, M.; Dumeignil, F. Towards the Sustainable Production of Acrolein by Glycerol Dehydration. *ChemSusChem* **2009**, *2*, 719–730. [CrossRef] [PubMed]

8. Chai, S.-H.; Wang, H.-P.; Liang, Y.; Xu, B.-Q. Sustainable production of acrolein: Gas-phase dehydration of glycerol over 12-tungstophosphoric acid supported on ZrO_2 and SiO_2. *Green Chem.* **2008**, *10*, 1087–1093. [CrossRef]

9. Wang, F.; Dubois, J.-L.; Ueda, W. Catalytic dehydration of glycerol over vanadium phosphate oxides in the presence of molecular oxygen. *J. Catal.* **2009**, *268*, 260–267. [CrossRef]

10. Ott, L.; Bicker, M.; Vogel, H. Catalytic dehydration of glycerol in sub- and supercritical water: A new chemical process for acrolein production. *Green Chem.* **2006**, *8*, 214–220. [CrossRef]

11. Haider, M.H.; Dummer, N.F.; Zhang, D.; Miedziak, P.; Davies, T.E.; Taylor, S.H.; Willock, D.J.; Knight, D.W.; Chadwick, D.; Hutchings, G.J. Rubidium- and caesium-doped silicotungstic acid catalysts supported on alumina for the catalytic dehydration of glycerol to acrolein. *J. Catal.* **2012**, *286*, 206–213. [CrossRef]

12. Magatani, Y.; Okumura, K.; Dubois, J.-L.; Devaux, J.-F. Catalyst and process for preparing acrolein and/or acrylic acid by dehydration reaction of glycerin. Patent WO2011033689A1, 24 March 2011.

13. Omata, K.; Matsumoto, K.; Murayama, T.; Ueda, W. Direct oxidative transformation of glycerol to acrylic acid over Nb-based complex metal oxide catalysts. *Catal. Today* **2016**, *259*, 205–212. [CrossRef]

14. Sato, S.; Akiyama, M.; Takahashi, R.; Hara, T.; Inui, K.; Yokota, M. Vapor-phase reaction of polyols over copper catalysts. *Appl. Catal. A Gen.* **2008**, *347*, 186–191. [CrossRef]

15. Chiu, C.-W.; Dasari, M.A.; Suppes, G.J.; Sutterlin, W.R. Dehydration of glycerol to acetol via catalytic reactive distillation. *AICHE J.* **2006**, *52*, 3543–3548. [CrossRef]

16. Sonnati, M.O.; Amigoni, S.; Taffin de Givenchy, E.P.; Darmanin, T.; Choulet, O.; Guittard, F. Glycerol carbonate as a versatile building block for tomorrow: Synthesis, reactivity, properties and applications. *Green Chem.* **2013**, *15*, 283–306. [CrossRef]

17. Behr, A.; Eilting, J.; Irawadi, K.; Leschinski, J.; Lindner, F. Improved utilisation of renewable resources: New important derivatives of glycerol. *Green Chem.* **2008**, *10*, 13–30. [CrossRef]

18. Tomishige, K.; Nakagawa, Y.; Tamura, M. Selective hydrogenolysis and hydrogenation using metal catalysts directly modified with metal oxide species. *Green Chem.* **2017**, *19*, 2876–2924. [CrossRef]

19. Bienholz, A.; Schwab, F.; Claus, P. Hydrogenolysis of glycerol over a highly active CuO/ZnO catalyst prepared by an oxalate gel method: Influence of solvent and reaction temperature on catalyst deactivation. *Green Chem.* **2010**, *12*, 290–295. [CrossRef]

20. Furikado, I.; Miyazawa, T.; Koso, S.; Shimao, A.; Kunimori, K.; Tomishige, K. Catalytic performance of Rh/SiO_2 in glycerol reaction under hydrogen. *Green Chem.* **2007**, *9*, 582–588. [CrossRef]

21. Yin, A.-Y.; Guo, X.-Y.; Dai, W.-L.; Fan, K.-N. The synthesis of propylene glycol and ethylene glycol from glycerol using Raney Ni as a versatile catalyst. *Green Chem.* **2009**, *11*, 1514–1516. [CrossRef]

22. Maglinao, R.L.; He, B.B. Catalytic Thermochemical Conversion of Glycerol to Simple and Polyhydric Alcohols Using Raney Nickel Catalyst. *Ind. Eng. Chem. Res.* **2011**, *50*, 6028–6033. [CrossRef]

23. Dasari, M.A.; Kiatsimkul, P.-P.; Sutterlin, W.R.; Suppes, G.J. Low-pressure hydrogenolysis of glycerol to propylene glycol. *Appl. Catal. A Gen.* **2005**, *281*, 225–231. [CrossRef]

24. Velasquez, M.; Santamaria, A.; Batiot-Dupeyrat, C. Selective conversion of glycerol to hydroxyacetone in gas phase over La_2CuO_4 catalyst. *Appl. Catal. B Environ.* **2014**, *160–161*, 606–613. [CrossRef]

25. Nimlos, M.R.; Blanksby, S.J.; Qian, X.; Himmel, M.E.; Johnson, D.K. Mechanisms of Glycerol Dehydration. *J. Phys. Chem. A* **2006**, *110*, 6145–6156. [CrossRef] [PubMed]

26. Antal, M.J.; Mok, W.S.L.; Roy, J.C.; Raissi, A.T.; Anderson, D.G.M. Pyrolytic sources of hydrocarbons from biomass. *J. Anal. Appl. Pyrolysis* **1985**, *8*, 291–303. [CrossRef]

27. Chai, S.-H.; Wang, H.-P.; Liang, Y.; Xu, B.-Q. Sustainable production of acrolein: Investigation of solid acid-base catalysts for gas-phase dehydration of glycerol. *Green Chem.* **2007**, *9*, 1130–1136. [CrossRef]

28. Haider, M.H.; Dummer, N.F.; Knight, D.W.; Jenkins, R.L.; Howard, M.; Moulijn, J.; Taylor, S.H.; Hutchings, G.J. Efficient green methanol synthesis from glycerol. *Nat. Chem.* **2015**, *7*, 1028–1032. [CrossRef] [PubMed]

29. Smith, L.R.; Smith, P.J.; Mugford, K.S.; Douthwaite, M.; Dummer, N.F.; Willock, D.J.; Howard, M.; Knight, D.W.; Taylor, S.H.; Hutchings, G.J. New insights for the valorisation of glycerol over MgO catalysts in the gas-phase. *Catal. Sci. Technol.* **2019**, *9*, 1464–1475. [CrossRef]

30. Chen, J.; Cai, Q.; Lu, L.; Leng, F.; Wang, S. Upgrading of the Acid-Rich Fraction of Bio-oil by Catalytic Hydrogenation-Esterification. *ACS Sustain. Chem. Eng.* **2017**, *5*, 1073–1081. [CrossRef]

31. Idriss, H.; Diagne, C.; Hindermann, J.P.; Kiennemann, A.; Barteau, M.A. Reactions of Acetaldehyde on CeO_2 and CeO_2-Supported Catalysts. *J. Catal.* **1995**, *155*, 219–237. [CrossRef]

32. Chen, T.L.; Mullins, D.R. Adsorption and Reaction of Acetaldehyde over $CeO_X(111)$ Thin Films. *J. Phys. Chem. C* **2011**, *115*, 3385–3392. [CrossRef]

33. Calaza, F.C.; Xu, Y.; Mullins, D.R.; Overbury, S.H. Oxygen Vacancy-Assisted Coupling and Enolization of Acetaldehyde on $CeO_2(111)$. *J. Am. Chem. Soc.* **2012**, *134*, 18034–18045. [CrossRef] [PubMed]

34. Mamontov, E.; Egami, T.; Brezny, R.; Koranne, M.; Tyagi, S. Lattice Defects and Oxygen Storage Capacity of Nanocrystalline Ceria and Ceria-Zirconia. *J. Phys. Chem. B* **2000**, *104*, 11110–11116. [CrossRef]
35. Qiao, Z.-A.; Wu, Z.; Dai, S. Shape-Controlled Ceria-based Nanostructures for Catalysis Applications. *ChemSusChem* **2013**, *6*, 1821–1833. [CrossRef]
36. McBride, J.R.; Hass, K.C.; Poindexter, B.D.; Weber, W.H. Raman and X-ray studies of $Ce_{1-x}RExO_{2-y}$, where RE = La, Pr, Nd, Eu, Gd, and Tb. *J. Appl. Phys.* **1994**, *76*, 2435–2441. [CrossRef]
37. Hua, G.; Zhang, L.; Fei, G.; Fang, M. Enhanced catalytic activity induced by defects in mesoporous ceria nanotubes. *J. Mater. Chem.* **2012**, *22*, 6851–6855. [CrossRef]
38. Nolan, M.; Parker, S.C.; Watson, G.W. Reduction of NO_2 on Ceria Surfaces. *J. Phys. Chem. B* **2006**, *110*, 2256–2262. [CrossRef]
39. Guo, M.; Lu, J.; Wu, Y.; Wang, Y.; Luo, M. UV and Visible Raman Studies of Oxygen Vacancies in Rare-Earth-Doped Ceria. *Langmuir* **2011**, *27*, 3872–3877. [CrossRef]

Article

Preparation of Solid Fuel Hydrochar over Hydrothermal Carbonization of Red Jujube Branch

Zhiyu Li [1,2], **Weiming Yi** [1,*], **Zhihe Li** [1], **Chunyan Tian** [1], **Peng Fu** [1], **Yuchun Zhang** [1], **Ling Zhou** [3,*] **and Jie Teng** [4,*]

[1] School of Agricultural Engineering and Food Science, Shandong University of Technology, Zibo 255000, China; lizhiyu@sdut.edu.cn (Z.L.); lizhihe@sdut.edu.cn (Z.L.); tiancy@sdut.edu.cn (C.T.); fupsklcc@126.com (P.F.); zhangyc@sdut.edu.cn (Y.Z.)

[2] Key Laboratory of Modern Agricultural Engineering, Department of Education of Xinjiang Uygur Autonomous Region, Alar 843300, China

[3] College of Mechanic and Electrical Engineering, Tarim University, Alar 843300, China

[4] School of Agricultural Sciences, Jiangxi Agricultural University, Nanchang 330045, China

* Correspondence: yiweiming@sdut.edu.cn (W.Y.); zhoul-007@163.com (L.Z.); tengjie33@163.com (J.T.); Tel.: +86-0533-2786-398 (W.Y.); +86-0997-4683-855 (L.Z.); +86-0791-8381-3185 (J.T.)

Received: 26 November 2019; Accepted: 14 January 2020; Published: 18 January 2020

Abstract: Biomass energy is becoming increasingly important, owing to the decreasing supply of fossil fuels and growing environmental problems. Hydrothermal carbonization (HTC) is a promising technology for producing solid biofuels from agricultural and forestry residues because of its lower fossil-fuel consumption. In this study, HTC was used to upgrade red jujube branch (RJB) to prepare hydrochar at six temperatures (220, 240, 260, 280, 300, and 320 °C) for 120 min, and at 300 °C for 30, 60, 90, and 120 min. The results showed that the energy recovery efficiency (ERE) reached maximum values of 80.42% and 79.86% at a residence time of 90 min and a reaction temperature of 220 °C, respectively. X-ray diffraction results and Fourier transform infrared spectroscopy measurements show that the microcrystal features of RJB were destroyed, whereas the hydrochar contained an amorphous structure and mainly lignin fractions at increased temperatures. Thermogravimetric analysis shows that the hydrochar had better fuel qualities than RJB, making hydrochar easier to burn.

Keywords: red jujube branch; hydrothermal carbonization; hydrochar; energy recovery efficiency; solid fuel

---

1. Introduction

Fossil-fuel combustion increases CO_2 emissions, leading to global warming. The development of renewable energies, such as biomass energy, has the potential to reduce the global dependence on fossil fuels [1]. With the development of the Xinjiang jujube industry in China, the output of red jujube branch (RJB) exceeded 1.9 million tons in 2012. RJB is the by-product of the development of the jujube industry. It has a high lignocellulose content, which is an excellent renewable biological resource. However, biomass disposal has become a contentious issue. When hydrochar is used as a solid fuel, it helps reduce NO_x emissions during the combustion of conventional energy sources [2]. Recently, hydrothermal carbonization (HTC) has helped increase bio-energy production efficiency, and HTC has received attention owing to its ability to convert organic waste into solid fuel [3–5]. The pathway to convert agricultural and forestry residues into solid fuels may be a viable RJB treatment method for reducing environmental pollution from RJB, while also providing new energy sources [6].

Compared with the biomass treatment technology of pyrolysis and combustion, HTC has the advantages of low energy input and mild reaction conditions [7]. Therefore, HTC is a good approach to convert biomass feedstocks to upgraded solid products or high-performance fuels. RJB has great potential as a raw material for the production of hydrochar [8]. Thus far, although RJB is a typical biomass in northwestern China, there have been few studies on the HTC of RJB for hydrochar production. Hydrochar can specifically be used as a solid fuel, activated carbon, and in other such applications. Presently, many studies focus on the HTC of other biomass [9–12].

In this study, the effects of modifying the reaction conditions of HTC on hydrochar production was examined. The results were analyzed to comprehensively evaluate the effect of temperature and time on product characteristics. The main objectives of this study are: (1) to report on the study of the structural and morphological properties of the hydrochar; (2) discuss the investigation of the combustion behavior of the hydrochar; and (3) report on the study of the fuel properties and combustion performance, as well as the evaluation of the optimal reaction conditions for the preparation of the hydrochar. The research results of RJB can provide readers with a deeper understanding of the HTC of RJB to prepare the hydrochar and provide a theoretical basis for the energy utilization of RJB. In terms of the efficient utilization of energy, the HTC of RJB seems to be a viable pathway.

2. Materials and Methods

2.1. Materials

The RJB used in the experiment was sourced from Alar City, Xinjiang Uygur Uyghur Autonomous Region. The industrial analysis of the red jujube branch is shown in Table 1.

Table 1. The physical and chemical properties of red jujube branch (RJB).

Sample	Moisture (%)	Proximate Analysis (%)			Heating Value (MJ kg^{-1})
		Fixed Carbon	**Volatile Matter**	**Ash**	
RJB	8.85	17.91	73.57	2.5	19.06

2.2. HTC Experiment

HTC was conducted in a laboratory-scale 500-mL high-pressure reactor using a magnetic stirrer (Parr 4575HP/HT, USA). The reactants were mixed to form a slurry, with RJB and distilled water at a ratio of 1:10 (*w/v*). The sample was stirred at a rate of 300 rpm throughput the experiment using a magnetic actuator with constant torque. Meanwhile, the heating rate was set at 10 °C/min. Then, the autoclave was pressurized to 4.0 MPa with nitrogen. The slurry used in each experiment was weighed and then poured into the reactor. The reactor was then sealed with a reactor cover and purged with nitrogen to remove the air. The control system was then turned on to set the temperature required for the reaction. The temperature inside the reactor was monitored continuously. The reaction residence time was calculated based on the instant the temperature of the reactor reached the required value. The effect of the reaction temperatures of 220, 240, 260, 280, 300, and 320 °C on the HTC of RJB was studied at a residence time of 120 min. In addition, the effect of the residence times of 30, 60, 90, and 120 min on the HTC of RJB was investigated at a reaction temperature of 300 °C. The control system was shut down after completion of each HTC experiment. After the reactor was rapidly cooled, the reactor cover was opened, and the reaction product was collected in a beaker. Finally, the weight of the collected product was recorded. The solid product was separated from the process water using a vacuum suction filter. The weights of the separated solid and liquid products were recorded. The solid and liquid yields were calculated using the ratio of the weight of the solid and liquid products, respectively, to the initial weight of the raw slurry. The gas yield was calculated using the difference [13]. Each test was conducted three times, and the average of the three tests was used as the test value. The energy recovery efficiency (ERE) was calculated as the ratio of the calorific value of dry hydrochar and raw materials to the mass yield.

$$\text{Solid yield} = \frac{\text{Weight of solid after filtration}}{\text{Weight of raw slurry}} \tag{1}$$

$$\text{Liquid yield} = \frac{\text{Weight of liquid after filtration}}{\text{Weight of raw slurry}} \tag{2}$$

$$\text{Mass yield} = \frac{\text{Weight of dry matter in hydrochar}}{\text{Weight of dry solid matter in raw material}} \tag{3}$$

$$\text{Energy recovery efficiency} = \frac{\text{Calorific value of hydrochar}}{\text{Calorific value of raw material}} \times \text{Mass yield} \tag{4}$$

2.3. Analytical Method

A powder sample with a particle size of 250 μm was prepared by sieving for testing. Proximate testing was performed on a YX-GYFX 7701 automatic industrial analyzer. The higher heating values of the sample were measured by a YX-ZR bomb calorimeter. The Fourier transform infrared spectroscopy (FT-IR) test of the sample was performed using a Thermo Fisher Nicolet iS10 FT-IR spectrometer. The tableting sample was prepared using dry potassium bromide to reduce the interference of water. Spectrum data were recorded over a wavenumber range of 4000 to 400 cm^{-1} at a resolution of 4 cm^{-1}. Thermogravimetric analysis (TGA) of RJB and hydrochar was performed using an STA 449 F5 Jupiter$^{\circledR}$ thermogravimetric analyzer. The data of the mass loss and mass loss rate of the sample were obtained under the analysis conditions shown in Table 2.

Table 2. Operating conditions for TGA.

TGA Conditions	Value
Sample weight	8 mg
Heating rate	10 $^{\circ}$C min^{-1}
Atmosphere	Air
Gas flow rate	30 mL min^{-1}
Starting temperature	30 $^{\circ}$C
Final holding temperature	800 $^{\circ}$C

3. Results and Discussion

3.1. Product Distribution

The effect of residence time on the three-state yield was studied. The three-state yield results are shown in Figure 1a–c. From 30 to 120 min, the solid yield showed an upward trend; however, both the liquid and gas yields showed downward trends. When the residence time reached 120 min, the solid yield reached a maximum value of 23.42%. The increase in residence time is mainly due to the decrease in solid yield. In the experimental group, longer residence times may allow more time for the hydrothermal reaction, resulting in an increased pore structure of hydrochar and the adsorption of more liquid, which may be the reason for the increase in the solid yield [13,14].

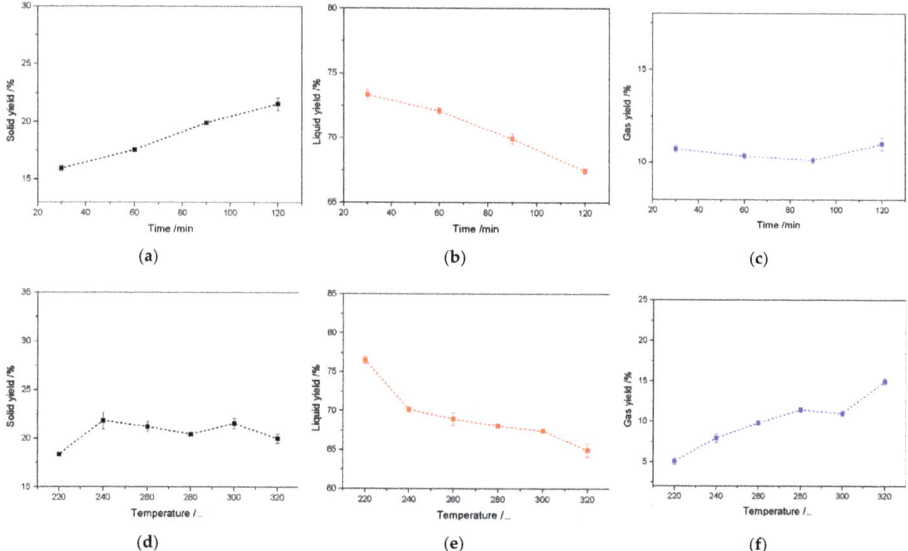

Figure 1. Effect of residence time (300 °C) and reaction temperature (120 min) on the yield of solid, liquid, and gas. (**a**) Solid yield; (**b**) liquid yield; (**c**) gas yield; (**d**) solid yield; (**e**) liquid yield; (**f**) gas yield.

Numerous studies have evaluated the HTC of RJB. Heilmann et al. studied the HTC of distiller's grains at an initial reaction pressure of 1.72 MPa. The results showed that the highest hydrochar yield was 45.6% at a temperature of 190 °C and residence time of 2 h. The lowest hydrochar yield was 30.2% at a temperature of 210 °C and residence time of 0.5 h [15]. Elaigwu et al. investigated the HTC of lignocellulosic waste material at a reaction pressure of 1.80 MPa. The results showed that the highest hydrochar yield was 37.87% at a temperature of 200 °C and residence time of 2 h. The lowest hydrochar yield was 30.6% at a temperature of 200 °C and residence time of 20 min [16]. The yield of our study was lower than those of reports, which is mainly attributed to the following three reasons. First, within the investigated parameters, longer residence times resulted in more complete carbonization, and consequently lower hydrochar yields [17]. Second, the reaction temperature significantly affected the physicochemical properties of the water (subcritical water) in the HTC process. The density of water changed up to one order of magnitude depending upon the temperature (0–350 °C), which allowed easier penetration of water in porous media and resulted in enhanced decomposition of the biomass [18]. Finally, oxygen was lost during the liquefaction reactions, which are parallel reactions to the hydrothermal process of the biomass [16].

The reaction temperature affected the HTC. The three-state yield results are shown in Figure 1b–f. With the increase in reaction temperature, the solid yield first increased, then decreased, and finally stabilized. With the increase in temperature from 220 to 320 °C, the liquid and gas yields showed downwards and upwards trends, respectively. The solid yield reached a maximum value of 21.82% at the reaction temperature of 240 °C; this may be due to the higher temperature causing the raw material to further decompose into a liquid phase, resulting in a lower solid yield. The reaction temperature, residence time, and material properties are the main factors in the HTC process [14].

3.2. Energy Recovery Efficiency

Energy was used in the HTC process to improve the fuel properties of the products [15]. Figure 2 shows the hydrochar ERE and heating value results. Therefore, the HTC process was evaluated by ERE, which was affected by the reduced product yield and increased heating value (Figures 1 and 2) [19]. The ERE of hydrochar after residence time treatment is shown in Figure 2a. The ERE trend is inconsistent with the solid yield over time. The ERE first increased and then decreased with increasing residence time. The ERE reached a maximum value of 80.42% for a residence time of 90 min.

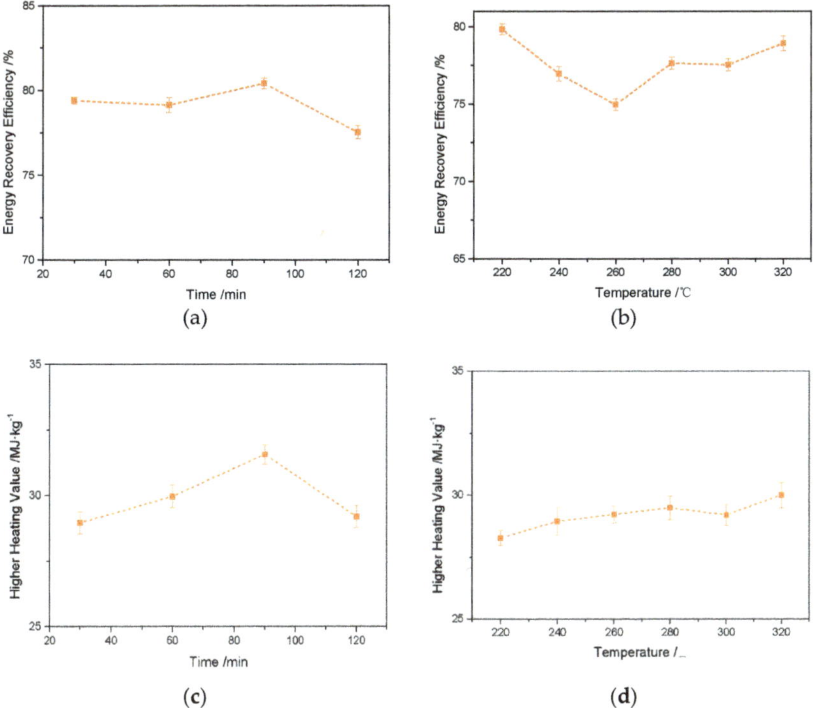

Figure 2. Energy recovery efficiency and higher heating value of hydrochar after hydrothermal carbonization (HTC) treatment (residence time (300 °C) and reaction temperature (120 min). (**a**) Residence time; (**b**) reaction temperature; (**c**) residence time; (**d**) reaction temperature.

As shown in Figure 2b, the effect of reaction temperature on ERE was studied. In contrast to the solid yield, the ERE was inconsistent with the reaction temperature. The ERE decreased and then increased with increasing reaction temperature. The ERE reached a maximum of 79.86% for the reaction temperature of 220 °C during the HTC process because the decarboxylation and dehydration reactions increased the temperature [19,20]. For lignocellulosic biomass, ERE generally decreased with increasing reaction temperature, which is consistent with the findings of Kambo [21].

The heating value of hydrochar increased with increasing reaction temperature. In addition, the solid yield was reduced owing to a decrease in the volatile-matter content caused by chemical dehydration and decarboxylation reactions (i.e., excluding CO_2) [19,22,23].

Therefore, the maximum ERE can be obtained under optimal temperature conditions during HTC [24–27]. ERE is an important factor in evaluating the production of hydrochar by HTC. The highest ERE is used to represent the optimum reaction temperature during HTC [28].

The effect of residence time and reaction temperature on the heating value of hydrochar is shown in Figure 2c,d. The results indicate that the heating value of hydrochar follows the order 90 min > 60 min > 120 min > 30 min, and the highest heating value is 31.57 MJ kg^{-1}. For identical treatment temperatures, the heating value of hydrochar increased with increasing residence time, but the heating value decreased more than 90 min. It is possible that the residence time was too long, which lead to an increase in the ash content and a decrease in the volatile content of the hydrochar. The heating value of hydrochar followed the order 320 °C > 280 °C > 260 °C > 300 °C > 240 °C > 220 °C, and the highest heating value was 30.00 MJ kg^{-1}. Similarly, for identical residence times, as the treatment temperature increased, the degree of removal of water or oxygen in the organic matter increased, resulting in an increase in the heating value of the hydrochar. The result shows that the optimum HTC temperature produced energy-rich solid fuels. Because lignin is the most abundant compound in RJB, HTC improves the heating value of lignin and the solid yield.

3.3. Characterization of RJB and Hydrochar

As shown in Figure 3a, the test results of the hydrochar obtained from the residence time and RJB contain broad peaks between 10° and 30°, which may be due to the diffraction of amorphous carbon [27,29]. This suggests that the RJB was carbonized. There are two crystalline peaks on RJB, which are the cellulosic peaks at 22.7° and 30.36°, respectively [30]. However, hydrochar has no crystalline peak at 30.36°, which may be due to the fact that hydrochar contained mainly amorphous components [31]. The results indicate that the crystalline structure of hydrochar was destroyed.

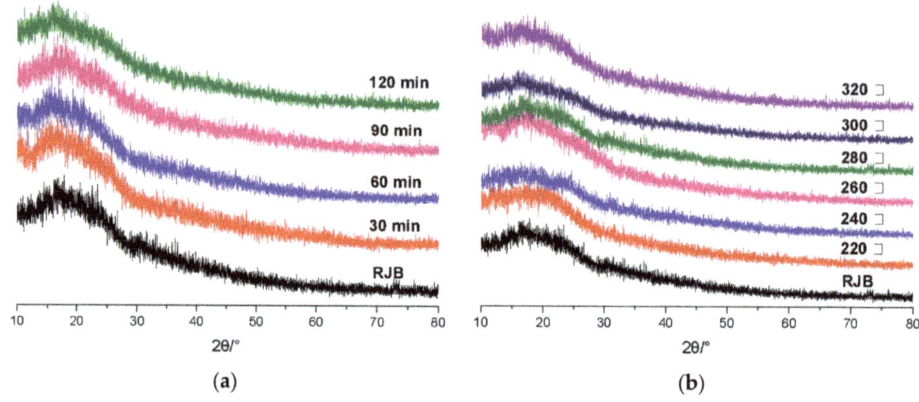

Figure 3. XRD spectra for RJB and hydrochar produced by HTC. (**a**) Residence time; (**b**) reaction temperature.

The XRD patterns for hydrochar obtained from the reaction temperature and RJB were compared, as shown in Figure 3b. There are two sharply crystalline cellulosic peaks from the cellulose crystal structure of RJB at 22.7° and 30.36°; however, there is no peak at 30.36° in the hydrochar, indicating that it contained mainly amorphous components. The decomposition of cellulose in RJB occurred during HTC, which realized the transformation of hydrochar from a microcrystalline structure to a non-crystalline structure. The main component of the hydrochar product obtained by HTC was lignin, which has an amorphous structure [32].

The FT-IR spectra of RJB and hydrochar are shown in Figure 4. The FT-IR spectra of hydrochar obtained from the residence time and reaction temperature are similar, differing only in the intensity of some peaks. The FT-IR spectrum of hydrochar was similar in part to that of RJB. The FT-IR spectra indicated that all samples contained an -OH functional group at 3500–3300 cm^{-1}.

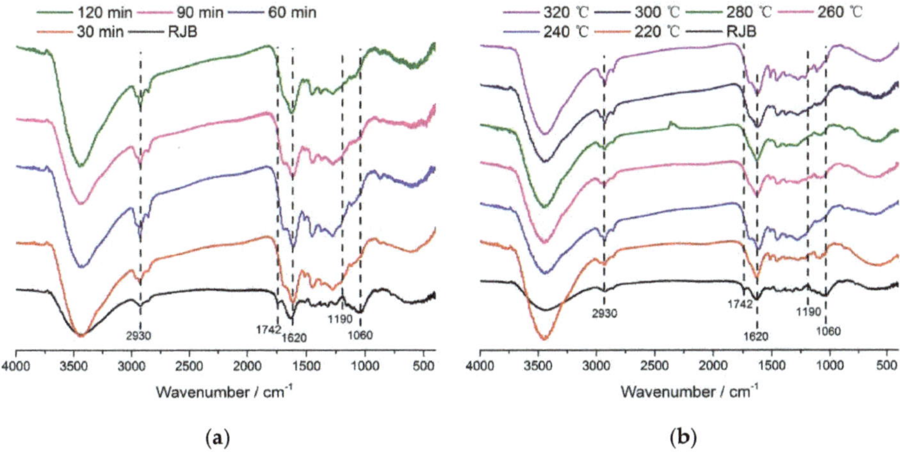

Figure 4. FT-IR spectra of RJB and its derived hydrochars. (**a**) Residence time; (**b**) reaction temperature.

The peak intensity exhibits no clear changes with increasing residence time (or reaction temperature), which confirms that -OH was not decomposed during HTC, as shown in Figure 4a or Figure 4b. The C-H stretching vibration at 3000–2800 cm^{-1} was attributed to either -CH_2 or -CH_3 functional groups, and the intensity of the peak indicates the presence of an aliphatic compound in the hydrochar.

The stretching vibration peak of aliphatic C-H obtained from the hydrochar tended to increase with residence time from 30 to 120 min (or from 220 to 320 °C; see Figure 4b), as shown in Figure 4a [33]. The peak at 1742 cm^{-1} is derived from the ester C=O stretching vibration of hemicelluloses and lignin [34].

The ester carbonyl group of the spectrum of RJB at 1734 cm^{-1} disappeared completely in the hydrochar, which may be because of the decomposition under hydrothermal conditions, including residence time or reaction temperature [35,36]. The peaks of RJB and all hydrochars around 1620 cm^{-1} correspond to the stretching vibration of the C=C of the aromatic groups in lignin [20], confirming the stability of the aromatic structure of lignin [33]. This indicates that lignin does not completely decompose under hydrothermal conditions [37].

D-xylose and cellulose have a C-O stretching vibration peak at 1000–1200 cm^{-1}, which means that there was either an ether bond or a methoxy group [20,37]. This -O peak decreased or disappeared with gradually increasing residence time (or reaction temperature), which may be attributed to the decomposition of carboxyl groups during the HTC process [20,31,37]. In addition, previous studies have shown that the C-O peak does not exist in the hydrochar components after HTC [20,33].

3.4. Combustion Behavior of Hydrochar

The thermogravimetry (TG) and derivative thermogravimetry (DTG) profiles of RJB hydrochar at different hydrothermal temperatures are shown in Figure 5. Most of the hydrochars (except the raw material) had a mass loss of more than 91% between 250 and 540 °C, and these hydrochars remained relatively stable below 250 °C. The raw material had a mass loss of 90% in the range of 200 to 550 °C. The maximum mass loss of the hydrochar occurred at approximately 460 °C, whereas the maximum mass loss of the raw material occurred at 319 °C. The results show that the weight-loss temperature of hydrochar moved to the high temperature zone after HTC, and HTC reduced the temperature range of the combustion zone. This is because the hydrothermal reaction destroys the original structure of the biomass, allowing some of the volatile components to dissolve in the water, producing a more stable aromatic carbon. In addition, the burnout temperature of the hydrochar is lower than that of the raw material, making the hydrochar easier to burn out, which is beneficial for use as a fuel [37].

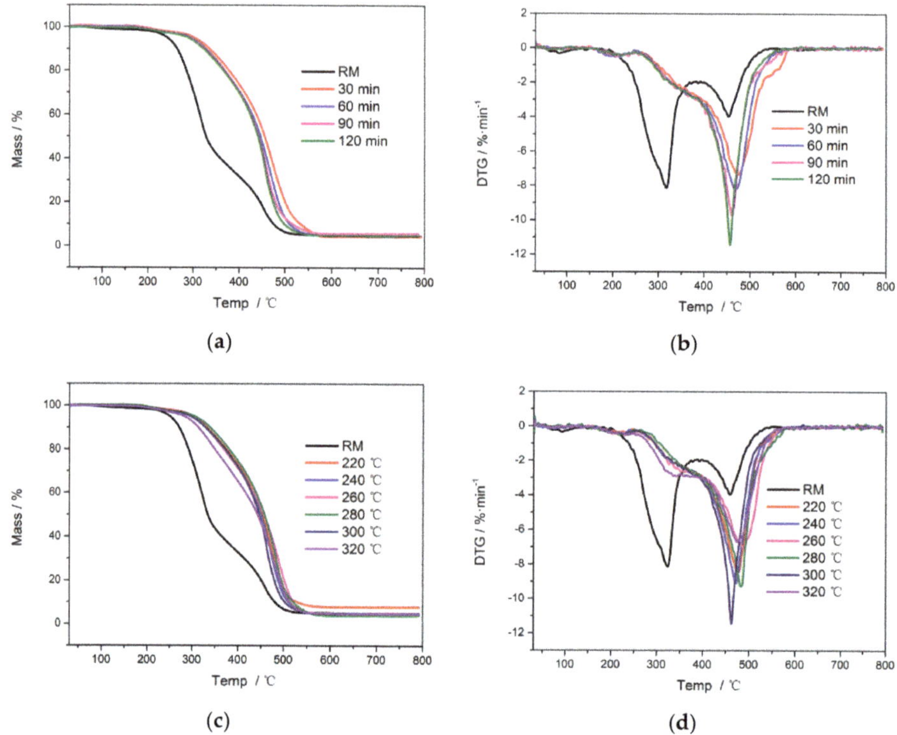

Figure 5. TGA of red jujube branch hydrochar. (**a**) TG of residence time; (**b**) derivative thermogravimetry (DTG) of residence time; (**c**) TG of reaction temperature; (**d**) DTG of reaction temperature. Note: The raw material is abbreviated as RM.

According to the peak of the remaining mass of the TG and the DTG curve, the thermostability of the treatment group was better than that of the raw material. The thermal stability increased with increasing residence time and reaction temperature. This may be because the volatile matter was reduced, and the fixed carbon was enhanced at higher temperature and longer residence times; these are the main factors affecting thermal stability [30]. HTC can cause many cracks in the hydrochar structures (Figure 5). Therefore, the combustion temperature of hydrochar is more concentrated [28].

The raw material is referred to as RM.

4. Conclusions

The fuel qualities of the produced hydrochars were evaluated by ERE. The results confirmed that RJB can produce hydrochar products with a higher ERE than RJB by HTC. ERE also increases with the increase in the severity of HTC reactions (longer time and higher temperature). Studies have found that the ERE of the hydrochar increased at higher reaction temperatures; however, most can be achieved at moderate temperatures. The experimental results showed that the ERE reached maximum values of 80.42% and 79.86% at a residence time of 90 min and a reaction temperature of 220 °C, respectively. The FT-IR and XRD measurements were used to obtain information about the chemical structure and thermal properties of the related hydrochar. The results showed that the microcrystal features of RJB were destroyed, and the hydrochar contained an amorphous structure and mainly lignin fractions at increased temperatures. TGA showed that the hydrochar has better fuel qualities than the RJB. For example, the burnout temperature and combustion temperature range of hydrochar decreased, making hydrochar easier to burn.

Author Contributions: Conceptualization, Z.L. (Zhiyu Li), W.Y., and J.T.; methodology, Z.L. (Zhiyu Li), Z.L. (Zhihe Li), and J.T.; validation, Z.L. (Zhiyu Li) and J.T.; experiment, Z.L. (Zhiyu Li), C.T., P.F., and Y.Z.; resources, L.Z.; writing-original draft preparation, Z.L. (Zhiyu Li), Z.L. (Zhihe Li), and L.Z.; writing and editing, W.Y., C.T., Z.L. (Zhihe Li), Y.Z., and J.T., and all authors reviewed the article. All authors have read and agreed to the published version of the manuscript.

Funding: This research was supported by the National Natural Science Foundation of China, China (Nos. 51536009), Taishan Scholars Program of Shandong Province, Ph.D. Program of Shandong University of Technology (4041-418033), and the National Natural Science Foundation of China (No. 41601601, 61501249), Open Project of Xinjiang Key Laboratory of Modern Agricultural Engineering (TDNG2020104), Scientific and Technological Research Projects in Key Fields of Xinjiang Production and Construction Corps (2019AB028), Youth Innovation Support Program of Shandong Colleges and Universities (2019KJD013).

Conflicts of Interest: The authors declare no conflict of interest.

References

1. Lynam, J.G.; Reza, M.T.; Yan, W.; Vásquez, V.R.; Coronella, C.J. Hydrothermal carbonization of various lignocellulosic biomass. *Biomass Convers. Biorefinery* **2015**, *5*, 173–181. [CrossRef]
2. Zheng, C.; Ma, X.; Yao, Z.; Chen, X. The properties and combustion behaviors of hydrochars derived from co-hydrothermal carbonization of sewage sludge and food waste. *Bioresour. Technol.* **2019**, *285*, 121347. [CrossRef] [PubMed]
3. Lucian, M.; Fiori, L. Hydrothermal carbonization of waste biomass: Process design, modeling, energy efficiency and cost analysis. *Energies* **2017**, *10*, 211. [CrossRef]
4. Basso, D.; Patuzzi, F.; Castello, D.; Baratieri, M.; Rada, E.C.; Weiss-Hortala, E.; Fiori, L. Agro-industrial waste to solid biofuel through hydrothermal carbonization. *Waste Manag.* **2016**, *47*, 114–121. [CrossRef]
5. Song, E.; Park, S.; Kim, H. Upgrading hydrothermal carbonization (HTC) hydrochar from sewage sludge. *Energies* **2019**, *12*, 2383. [CrossRef]
6. Laginhas, C.; Nabais, J.M.V.; Titirici, M.M. Activated carbons with high nitrogen content by a combination of hydrothermal carbonization with activation. *Microporous. Mesoporous. Mater.* **2016**, *226*, 125–132. [CrossRef]
7. Lin, Y.; Ma, X.; Peng, X.; Yu, Z. A mechanism study on hydrothermal carbonization of waste textile. *Energy Fuels* **2016**, *30*, 7746–7754. [CrossRef]
8. Yao, Z.; Ma, X. Hydrothermal carbonization of Chinese fan palm. *Bioresour. Technol.* **2019**, *282*, 28–36. [CrossRef]
9. Parshetti, G.K.; Liu, Z.; Jain, A.; Srinivasan, M.P.; Balasubramanian, R. Hydrothermal carbonization of sewage sludge for energy production with coal. *Fuel* **2013**, *111*, 201–210. [CrossRef]
10. Yao, Z.; Ma, X.; Lin, Y. Effects of hydrothermal treatment temperature and residence time on characteristics and combustion behaviors of green waste. *Appl. Therm. Eng.* **2016**, *104*, 678–686. [CrossRef]
11. Yao, Z.; Ma, X.; Wang, Z.; Chen, L. Characteristics of co-combustion and kinetic study on hydrochar with oil shale: A thermogravimetric analysis. *Appl. Therm. Eng.* **2017**, *110*, 1420–1427. [CrossRef]
12. Chen, X.; Ma, X.; Peng, X.; Lin, Y.; Yao, Z. Conversion of sweet potato waste to solid fuel via hydrothermal carbonization. *Bioresour. Technol.* **2018**, *249*, 900–907. [CrossRef] [PubMed]
13. Bhatt, D.; Shrestha, A.; Dahal, R.; Acharya, B.; Basu, P.; MacEwen, R. Hydrothermal carbonization of biosolids from waste water treatment plant. *Energies* **2018**, *11*, 2286. [CrossRef]
14. Nizamuddin, S.; Baloch, H.A.; Griffin, G.J.; Mubarak, N.M.; Bhutto, A.W.; Abro, R.; Mazari, S.A.; Ali, B.S. An overview of effect of process parameters on hydrothermal carbonization of biomass. *Renew. Sustain. Energy Rev.* **2017**, *73*, 1289–1299. [CrossRef]
15. Heilmann, S.M.; Jader, L.R.; Sadowsky, M.J.; Schendel, F.J.; Keitz, M.G.V.; Valentas, K.J. Hydrothermal carbonization of distiller's grains. *Biomass Bioenergy* **2011**, *35*, 2526–2533. [CrossRef]
16. Elaigwu, S.E.; Greenway, G.M. Microwave-assisted and conventional hydrothermal carbonization of lignocellulosic waste material: Comparison of the chemical and structural properties of the hydrochars. *J. Anal. Appl. Pyrolysis* **2016**, *118*, 1–8. [CrossRef]

17. Sermyagina, E.; Saari, J.; Kaikko, J.; Vakkilainen, E. Hydrothermal carbonization of coniferous biomass: Effect of process parameters on mass and energy yields. *J. Anal. Appl. Pyrolysis* **2015**, *113*, 551–556. [CrossRef]
18. Kambo, H.S.; Dutta, A. Comparative evaluation of torrefaction and hydrothermal carbonization of lignocellulosic biomass for the production of solid biofuel. *Energy Convers. Manag.* **2015**, *105*, 746–755. [CrossRef]
19. Kim, D.; Lee, K.; Park, K.Y. Hydrothermal carbonization of anaerobically digested sludge for solid fuel production and energy recovery. *Fuel* **2014**, *130*, 120–125. [CrossRef]
20. Kim, D.; Lee, K.Y.; Park, K.Y. Upgrading the characteristics of biochar from cellulose, lignin, and xylan for solid biofuel production from biomass by hydrothermal carbonization. *J. Ind. Eng. Chem.* **2016**, *42*, 95–100. [CrossRef]
21. Kambo, H.S. Energy Densification of Lignocellulosic Biomass via Hydrothermal Carbonization and Torrefaction. Doctoral Dissertation, University of Guelph, Guelph, ON, Canada, 2014.
22. Byrappa, K.; Adschiri, T. Hydrothermal technology for nanotechnology. *Prog. Cryst. Growth Charact. Mater.* **2007**, *53*, 117–166. [CrossRef]
23. Funke, A. Hydrothermal carbonization of biomass: A summary and discussion of chemical mechanisms for process engineering. *Biofuels Bioprod. Biorefining* **2010**, *4*, 160–177. [CrossRef]
24. Liu, Z.; Balasubramanian, R. Upgrading of waste biomass by hydrothermal carbonization (HTC) and low temperature pyrolysis (LTP): A comparative evaluation. *Appl. Energy* **2014**, *114*, 857–864. [CrossRef]
25. Berge, N.D.; Ro, K.S.; Mao, J.; Flora, J.R.V.; Chappell, M.A.; Bae, S. Hydrothermal carbonization of municipal waste streams. *Environ. Sci. Technol.* **2011**, *45*, 5696–5703. [CrossRef] [PubMed]
26. Lu, X.; Pellechia, P.J.; Flora, J.R.; Berge, N.D. Influence of reaction time and temperature on product formation and characteristics associated with the hydrothermal carbonization of cellulose. *Bioresour. Technol.* **2013**, *138*, 180–190. [CrossRef]
27. Kang, S.; Li, X.; Fan, J.; Chang, J. Characterization of hydrochars produced by hydrothermal carbonization of lignin, cellulose, D-xylose, and wood meal. *Ind. Eng. Chem. Res.* **2012**, *51*, 9023–9031. [CrossRef]
28. Kim, D.; Yoshikawa, K.; Park, K. Characteristics of biochar obtained by hydrothermal carbonization of cellulose for renewable energy. *Energies* **2015**, *8*, 14040–14048. [CrossRef]
29. Kumar, S.; Loganathan, V.A.; Gupta, R.B.; Barnett, M.O. An assessment of U (VI) removal from groundwater using biochar produced from hydrothermal carbonization. *J. Environ. Manag.* **2011**, *92*, 2504–2512. [CrossRef]
30. Liu, C.; Huang, X.; Kong, L. Efficient low temperature hydrothermal carbonization of Chinese reed for biochar with high energy density. *Energies* **2017**, *10*, 2094. [CrossRef]
31. Song, C.; Zheng, H.; Shan, S.; Wu, S.; Wang, H.; Christie, P. Low-temperature hydrothermal carbonization of fresh pig manure: Effects of temperature on characteristics of hydrochars. *J. Environ. Eng.* **2019**, *145*, 04019029. [CrossRef]
32. Jayaramudu, J.; Agwuncha, S.; Ray, S.S.; Rajulu, A.V. Studies on the chemical resistance and mechnical properties of natural polyalthia cerasoides woven fabric/glass hybridized epoxy composites. *Adv. Mater. Lett.* **2015**, *6*, 114–119. [CrossRef]
33. Janković, B.; Manić, N.; Dodevski, V.; Radović, I.; Pijović, M.; Katnić, Đ.; Tasić, G. Physico-chemical characterization of carbonized apricot kernel shell as precursor for activated carbon preparation in clean technology utilization. *J. Clean. Prod.* **2019**, *236*, 117614. [CrossRef]
34. Sivasankarapillai, G.; McDonald, A.G. Synthesis and properties of lignin-highly branched poly (ester-amine) polymeric systems. *Biomass Bioenergy* **2011**, *35*, 919–931. [CrossRef]
35. Reza, M.T.; Uddin, M.H.; Lynam, J.G.; Hoekman, S.K.; Coronella, C.J. Hydrothermal carbonization of loblolly pine: Reaction chemistry and water balance. *Biomass Convers. Biorefinery* **2014**, *4*, 311–321. [CrossRef]
36. Liu, Z.; Quek, A.; Hoekman, S.K.; Balasubramanian, R. Production of solid biochar fuel from waste biomass by hydrothermal carbonization. *Fuel* **2013**, *103*, 943–949. [CrossRef]
37. Zhou, S.; Han, L.; Yang, Z.; Ma, Q. Influence of hydrothermal carbonization temperature on combustion characteristics of livestock and poultry manures. *Trans. Chin. Soc. Agric. Eng.* **2017**, *33*, 233–240.

MDPI
St. Alban-Anlage 66
4052 Basel
Switzerland
Tel. +41 61 683 77 34
Fax +41 61 302 89 18
www.mdpi.com

Energies Editorial Office
E-mail: energies@mdpi.com
www.mdpi.com/journal/energies